COMMERCIAL AND PRE-COMMERCIAL CELL DETECTION TECHNOLOGIES FOR DEFENCE AGAINST BIOTERROR

NATO Science for Peace and Security Series

This Series presents the results of scientific meetings supported under the NATO Programme: Science for Peace and Security (SPS).

The NATO SPS Programme supports meetings in the following Key Priority areas: (1) Defence Against Terrorism; (2) Countering other Threats to Security and (3) NATO, Partner and Mediterranean Dialogue Country Priorities. The types of meeting supported are generally "Advanced Study Institutes" and "Advanced Research Workshops". The NATO SPS Series collects together the results of these meetings. The meetings are co-organized by scientists from NATO countries and scientists from NATO's "Partner" or "Mediterranean Dialogue" countries. The observations and recommendations made at the meetings, as well as the contents of the volumes in the Series, reflect those of participants and contributors only; they should not necessarily be regarded as reflecting NATO views or policy.

Advanced Study Institutes (ASI) are high-level tutorial courses to convey the latest developments in a subject to an advanced-level audience.

Advanced Research Workshops (ARW) are expert meetings where an intense but informal exchange of views at the frontiers of a subject aims at identifying directions for future action.

Following a transformation of the programme in 2006 the Series has been re-named and re-organised. Recent volumes on topics not related to security, which result from meetings supported under the programme earlier, may be found in the NATO Science Series.

The Series is published by IOS Press, Amsterdam, and Springer Science and Business Media, Dordrecht, in conjunction with the NATO Public Diplomacy Division.

Sub-Series

A.	Chemistry and Biology	Springer Science and Business Media
B.	Physics and Biophysics	Springer Science and Business Media
C.	Environmental Security	Springer Science and Business Media
D.	Information and Communication Security	IOS Press
E.	Human and Societal Dynamics	IOS Press

http://www.nato.int/science
http://www.springer.com
http://www.iospress.nl

Sub-Series E: Human and Societal Dynamics – Vol. 39 ISSN 1874-6276

Commercial and Pre-Commercial Cell Detection Technologies for Defence against Bioterror

Technology, Market and Society

Edited by

Laura M. Lechuga

Microelectronics National Center (IMM-CNM), CSIC, Spain

Fred P. Milanovich

Lawrence Livermore National Laboratory, Livermore, USA

Petr Skládal

Masaryk University, Czech Republic

Oleg Ignatov

Institute of Biochemistry & Physiology of Plants & Microorganisms, Russian Academy of Sciences, Russia

and

Thomas R. Austin

The Boeing Company, USA

IOS Press

Amsterdam • Berlin • Oxford • Tokyo • Washington, DC

Published in cooperation with NATO Public Diplomacy Division

Proceedings of the NATO Advanced Research Workshop on Commercial and Pre-Commercial Cell Detection Technologies for Defence against Bioterror – Technology, Market and Society
Brno, Czech Republic
3–6 September 2006

ISBN 978-1-58603-858-8
Library of Congress Control Number: 2008925577

Publisher
IOS Press
Nieuwe Hemweg 6B
1013 BG Amsterdam
Netherlands
fax: +31 20 687 0019
e-mail: order@iospress.nl

Distributor in the UK and Ireland
Gazelle Books Services Ltd.
White Cross Mills
Hightown
Lancaster LA1 4XS
United Kingdom
fax: +44 1524 63232
e-mail: sales@gazellebooks.co.uk

Distributor in the USA and Canada
IOS Press, Inc.
4502 Rachael Manor Drive
Fairfax, VA 22032
USA
fax: +1 703 323 3668
e-mail: iosbooks@iospress.com

PRINTED IN THE NETHERLANDS

L.M. Lechuga et al. (Eds.)
IOS Press, 2008

PREFACE

Bioterrorism and biological warfare employ living agents or toxins that can be disseminated or delivered by infected individuals, insects, aerosols, and by the contamination of water and food supplies. Biological warfare agents (BWA) such as bacteria, viruses, toxins and genetically engineered species are usually characterized as invisible, odor- and taste-free. Most biological agents can be thousands of times more lethal per unit than the most lethal chemical warfare agents. Unlike chemical agents, biological agents attack people stealthily with no observable reaction until after an incubation period (1–14 days). Current disease surveillance and response systems rely on post-symptomatic reporting. However, many infectious agents such as smallpox have a long latency to clinical symptoms, thereby eluding early detection and resulting in widespread, uncontrolled contagion. Consequently, the threat of deliberate dissemination of biological agents is the most complicated and problematic of the weapons of mass destruction facing mankind today.

This volume contains papers presented at the NATO Advanced Research Workshop "Commercial and Pre-commercial Cell Detection Technologies for Defense against Bioterror – Technology, Market and Society", held in Brno, Czech Republic in September 2006. As a response to the rapidly emerging threat of bioterrorism, the objectives of the workshop were: (i) to exchange information on commercially available technologies and equipment for defense against bioterrorism; (ii) to further the development of new biosensor system prototypes into a commercially available apparatus; and to explore human factors in BWA biosensors. During the Workshop the new commercial and pre-commercial technologies that are currently emerging in the world were presented and explained. On the other hand, we discussed the interaction of modern detection systems with society and we tried to improve the relation between the scientific community and commercial entities. As a summary, the major areas of activity during the Workshop were the following:

1) A presentation of the most advanced biosensors and biodetection system which can be found in the market or are quite close to commercialization. Systems as the BIOHAWK™, SASS 2000, RAPTOR, Bionas® 2500, OWLS, or a portable SPR were presented.
2) A presentation of the advances in the research of biodetection devices as DNA and protein microchips, micro and nanophotonic sensors, CMOS microsensor chips, electrochemical arrays, physical platforms, electro optical detection, mass detection, etc.
3) A description of the latest developments in the employment of bioreceptor layers for the selective detection of BWA, as protein signatures, molecular imprinted polymers, membrane engineering (MIME), cell signatures, monoclonal antibodies, synthetic antibodies, lytic phages, among others.
4) A deep discussion of the human factor: societal issues related to sensor development and employment for BWA detection.

The editors: Laura M. Lechuga, Fred P. Milanovich, Petr Skládal, Oleg Ignatov and Thomas R. Austin.

LIST OF CONTRIBUTORS

STEVEN KORNGUTH
The University of Texas at Austin
Institute Advanced Technology
512 232 4485
steve_kornguth@iat.utexas.edu
3925 West Braker Lane, Suitre 400
Austin, Texas 78759-5316
USA

LAURA M. LECHUGA
Biosensors Group
Microelectronics National Center (CNM), CSIC
+3491 8060789
laura@imm.cnm.csic.es
C/ Isaac Newton, 8 (PTM)
28760 Tres Cantos (Madrid)
SPAIN

PETR SKLADAL
Department of Biochemistry
Masaryk University
00420 5 4949 7010
skladal@chemi.muni.cz
Kotlarska 2, 611 37 Brno
CZECH REPUBLIC

MARCO MASCINI
University of Firenze
Dept of Chemistry
554573283
marco.mascini@unifi.it
Via della Lastruccia 3 50019
ITALY

DMITRI IVNITSKI
University of New Mexico
505-277-2563
ivnitski@unm.edu
MSC01 1120, 209 Farris Engineering Center
Albuquerque, NM 87131-0001
USA

OLEG IGNATOV
Institute of Biochemistry & Physiology of Plants & Microorganisms RAS
7(8452)970404
olegignatov@yahoo.com
Entuziastov av., 13 Saratov
410049 Russia
RUSSIA

VICTOR BUNIN
BIOTRONIX GmbH
bunin@biotronix.de
Neuendorfstr. 24a 16761 Henningsdorf,
Germany

FRANZ L. DICKERT
Institute of Analytical Chemistry and Food Chemistry
Vienna University
0043-1-4277-52317/52301
Franz.Dickert@univie.ac.at
Waehringer Str. 38
A-1090 Vienna
AUSTRIA

WERNER BAUMANN
Biophysics Dept.
University of Rostock
+49 381 498 6029
werner.baumann@uni-rostock.de
Gertrudenstr. 11A; D-18057 Rostock
GERMANY

RALF EHRET
BIONAS GmbH
+49 381 5196 243
ralf.ehret@bionas.de
Friedrich-Barnewitz-Straße 3
D-18119 Rostock
GERMANY

ANATOLY RESHETILOV
G. K. Skryabin Institute of Biochemistry and Physiology of Microorganisms - RAS
007(4967)73-16-66
anatol@ibpm.pushchino.ru
142290, Pushchino, Nauki Av., 5. Moscow region
RUSSIA

SPIRIDON KINTZIOS
Agricultural University of Athens
Lab. Plant Physiology
+302105294292
skin@aua.gr
Iera Odos 75, 11855 Athens
GREECE

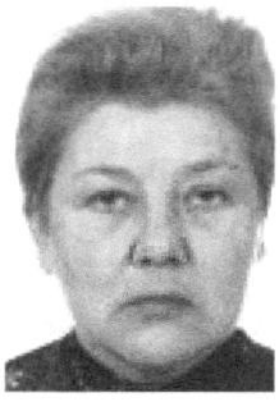

IRINA MOSKALENKO
Russian Research Center "Kurchatov Institute"
(495) 196-7617
ivm@nfi.kiae.ru
Kurchatov sq., 1
123182 Moscow
RUSSIA

FRED MILANOVICH
Lawrence Livermore National Laboratory
(925) 422-6838
milanovich1@llnl.gov
7000 East Ave, L-210, Livermore, CA 94550
USA

VALDAS LAURINAVICIUS
Institute of Biochemistry
+370 5 272 91 44
valdasL@bchi.lt
Mokslininku str. 12, Vilnius LT-08662
LITHUANIA

JAN KREJČÍ
BVT Technologies a.s.
420.541.513.545
info@bvt.cz
Hudcova 78c
612 00 Brno
CZECH REPUBLIC

Contents

Commercial and Pre-Commercial Cell Detection Technologies for Defence against Bioterror
L.M. Lechuga et al. (Eds.)
IOS Press, 2008

Societal Issues and Deployment of Integrated Biological Sensors

Ariane BECK and Steve KORNGUTH[1]
The Institute for Advanced Technology at The University of Texas at Austin, Austin, Texas

Abstract. The current work addresses the technical and social issues that must be considered when assessing the probability of infectious disease occurrence of natural or man-made origins, within a nation state or across the globe. The technical aspects of the problem relate to the ability to detect potentially subtle changes in the level of threat agents in an environment, the susceptibility of the host, and environmental conditions rendering the host vulnerable to infection. In order to determine an increase in disease emergence, it is necessary to establish a baseline of normal disease incidence in a population, the concentration of pathogens in each environmental region, and the presence of non-pathogenic organisms that are genomically or proteomically similar to pathogens. Once baselines are established for these factors, multiplexed sensors and data fusion technologies can be used to relate changes in the concentration of agents or in host susceptibility to the emergence of clinical symptoms in a community. One premise of this paper is that emergent disease is neither a function of infectious agents alone nor of susceptibility of the human or livestock target alone, but rather it is a probabilistic event dependent on multiple factors including the interaction between the pathogen and the host target. The probabilistic assessment of emergent disease requires large databases, deployed sensors, data fusion, and autonomous rapid decision making capabilities. The extensive deployment of pervasive surveillance systems can cause societal concerns regarding the balance between assuring wellness in a population while simultaneously respecting the privacy of individuals. Because emergent pandemic disease is a relatively low probability event, it is probable that a significant time lapse will occur between the gathering of information from the distributed sensors and the actual realization of pandemic disease. During this interval, societal perceptions may instigate public concern over compromised privacy, because benefits from the sensor deployment may not yet be realized. Possible adverse affects include changes in insurance rates of individuals, loss of employment opportunities, and other unanticipated negative consequences resulting from widespread data acquisition. The goal, however, of maintaining the wellness of society as a whole will require thoughtful balance between the potential loss of individual privacy and maintaining the wellness of the community.

Keywords. Biological agent, probabilistic assessment, biosensor, ethics, privacy rights, economic cost

Introduction

Intentional release of biological and chemical agents is a low probability, high consequence occurrence. This paper will introduce the basic approaches for

[1] Corresponding author: Steve Korngut. The Institute for Advanced Technology at The University of Texas at Austin, Austin

determining the presence of biological agents in the environment and of assessing host susceptibility to the threat. Advances in decoding the human genome and the genome of infectious agents, as well as proteomic markers on pathogens and toxins, have enabled the construction of probes that can bind agents with specificity and reasonable affinity. When coupled with current photonic and electronic transduction platforms, it has become possible to develop multiplexed detection and sensor systems that can quantify the levels of agent in an aerosol or aqueous environment, as well as assess host susceptibility to agents using blood samples. Multiplexed systems can reduce the probability of false positive or false negative responses, thereby increasing the reliability of the detection and identification system, which is defined as a reliable "probabilistic assessment tool" in this report. Such multiplexed sensor systems will have utility for both public and personal monitoring of infection of the individual or cohort of individuals examined by the sensor devices. The use of personal sensors in the home or cell phones raises potential concerns of compromised personal privacy. It is the intent of this paper to develop an approach by which societal institutions can establish an understanding of the trade-offs between ceding some level of privacy with an end goal of protecting public welfare. The recognition that intentional release of biological threat agents has thus far been a very infrequent but high impact event increases the complexity of privacy issues, as will be discussed.

1. Threat Assessment

1.1 Databases

Databases required for developing a reliable probabilistic assessment of biological threats include genetic, proteomic, and population criteria. Some databases focus on critical genomic sequences of pathogenic agents, or genomic sequences of the host that may confer susceptibility or resistance. Additionally, critical proteomic data have been catalogued relating to antigenic displays on the pathogen, and the altered expression of proteins in humans or livestock indicating exposure of the host to infectious agents. On a population scale, databases document normal disease incidence and steady state levels of pathogens in an environment, as well as markers of population wellness. Bacterial pathogens possess genomic pathogenicity islands encoding protein "factors" that affect adhesion of the bacterium to host tissues or the concentration of essential nutrients required for the growth of the pathogen. These islands may also code for the production of toxins involved in the generation of the disease state. More than fifty of these pathogenicity islands or factors have been characterized and can be detected by genomic screening methodologies. Pathogenicity islands are frequently transmitted via horizontal transfer of DNA between individual bacteria. Viruses, which do not have the ability to directly exchange DNA, do not contain pathogenicity islands, but rather have factors that are predictive of the degree of pathogenicity of the viral organism within a host species. The genes coding for the neuraminidase and hemagglutinin protein components of the influenza virus are examples of such pathogenicity related factors.

In addition to databases of genomic and proteomic markers of biological agents, it is important to map steady state levels of threat agents in many different regions of the world, because levels of threat agents vary geographically and across microenvironments. As an example, Bacillus anthracis spores or their antigenic determinants are present at much higher concentrations in soils near farms with large numbers of cattle than in urban areas. Additionally, Avian influenza virus levels in an environment where ducks, chickens, and geese are raised in densely concentrated facilities will be higher than those in other regions. Databases reflecting differences in regional levels of agents are therefore required.

Databases that identify host biomarkers correlated with increased or decreased susceptibility to infectious disease and toxins are the final element of background information needed for reliably assessing probability of disease occurrence. The human or livestock host will have allelic variability resulting in particular genetic codes that increase or decrease susceptibility to infectious agents. Included in this group are major histocompatibility markers and immune response modifiers, such as the interleukins, the detoxifying enzymes cytochrome P-450s, and glutathione-S-transferases. Susceptibility factors within a population and geographical region vary. Elderly populations or immune compromised patients, such as cancer patients treated with chemotherapy or radiation, AIDS patients, or organ transplant recipients, have increased susceptibility to disease. Such populations may serve as a reservoir for pathogenic organisms, thereby increasing the risk to the entire community. Examples of additional causes of increased susceptibility include concurrent infections in the population, large immigrations of infected populations, and floods that increase exposure to water borne parasites or cause contamination of potable water sources. Prior exposure of the host to infectious agents may also generate an immune response that can protect the host from developing clinical disease. Databases regarding each one of these factors are essential for determining population wellness in each region of interest.

1.2 Integrated Multiplexed Sensors

Once databases have been established, it becomes possible to deploy sensors that will be able to detect changes in the concentration of threat agents in the environment or of increased susceptibility of hosts to the agent. Multiplexed sensor platforms provide a distinct advantage for development of reliable platforms with low false positive and false negative responses. The multiplexed platform utilizes multiple probes for each signature element of a pathogen (a signature element could be an antigenic marker or a 22 base long polynucleotide sequence of a pathogenicity island). If three or four probes react with a target in the environment, the probability that the agent of concern is present is much higher than if only one of the probes reacts. Thus, increased resolution of threat probability is accomplished via the cross-platform design. Simultaneous testing with probes designed against markers of non-pathogenic microorganisms can serve as a further buffer against false positives; if these probes also react, it is likely that non-specific binding of probes to materials in the environment has occurred or that the sensor surface has become fouled and lost specificity. This type of fused data can serve to reduce the incidence of false positives, thereby reducing false alarms. Multiplexed systems also reduce the ability of an adversary to intentionally trigger an alarm by introducing a small but non-infectious region of pathogen into air or food samples. Such an act of sabotage can cause extensive economic dislocation to nations. An additional advantage of multiplexed biosensor platforms is the ability of the existing platform to adapt and rapidly respond to emergent pathogens or new threats.

1.3 Data Fusion and Assessment

Once acquired, the data must be converted into information via autonomous means and presented in a timely, readily understood iconographic format, as illustrated in Figure 1. Autonomous means of processing biosensor data eliminate errors and time delays due to loss of vigilance, ambiguity, fatigue, etc. experienced by a person in the loop. Techniques for correlating and processing data into information include data mining and fusion and utilizing weighting factors (i.e. intelligent agents, source credibility, and data re-fusion). Weighting factors based on the credibility of the source take into consideration that some sensors will be more reliable and/or accurate than others. A reliable integrated system requires employing intelligent agents (i.e. artificial intelligence) that can assess the historical accuracy, sensitivity, and specificity of each sensor element and integrate that information into a weighting mechanism. Achieving these capabilities requires a number of technologies currently in various stages of development.

Real world operations then act on the acquired intelligence by characterizing the biological substrate (i.e. the genetics, sex, susceptibility, and age of the host) and utilizing training, experience, and doctrine to respond appropriately to the threat, as shown in Figure 2. If assessment of the threat leads to the conclusion that there is a significant increase in the level of biological agent, rapid decision making is required. Immediate actions include deployment of vaccines or treatments, isolation of individuals with contagious disease, HAZMAT readiness, and preparation of hospitals, treatment facilities, or mortuaries. Military forces may be required to enforce perimeters for containment in either naturally occurring or intentional threats. Additionally, in the case of intentional threats, military action may be appropriate for the prevention of or defense from future attacks, as well as elements of retribution. In all of these situations, an effective response and positive outcome depends on agile probabilistic assessment of the situation. The paradigm promoted here is a systems approach to controlling emergent disease.

1.4 Sensors in the Public and Private Domain

Probabilistic assessment of biological threats requires broad distribution of sensors in both the public and private domain in order to collect sufficient data, as illustrated in Figure 3. There are several current technologies that would have a dominant role in monitoring the dispersion of threat agents in a community. Some of these platforms are area monitors that would present less risk of loss of personal privacy, while others would effectively be personal sensors that could compromise personal privacy. Examples of the former include biological agent sensor platforms embedded in public meeting arenas, in airplane cabins, on the external surface of buses, or in toilets in hospital emergency rooms. Examples of the latter could include cell phones with integrated sensor platforms, sensors embedded in toilets within the home, or dedicated handheld sensors for home monitoring. The components of the sensor platforms include a sample capture device, materials that bind threat agent components, transducers that report a binding event between a threat agent and the material on the platform, and a transmitter reporting the results to some central data collector. The include a sample capture device, materials that bind threat agent components, transducers that report a binding event between a threat agent and the material on the

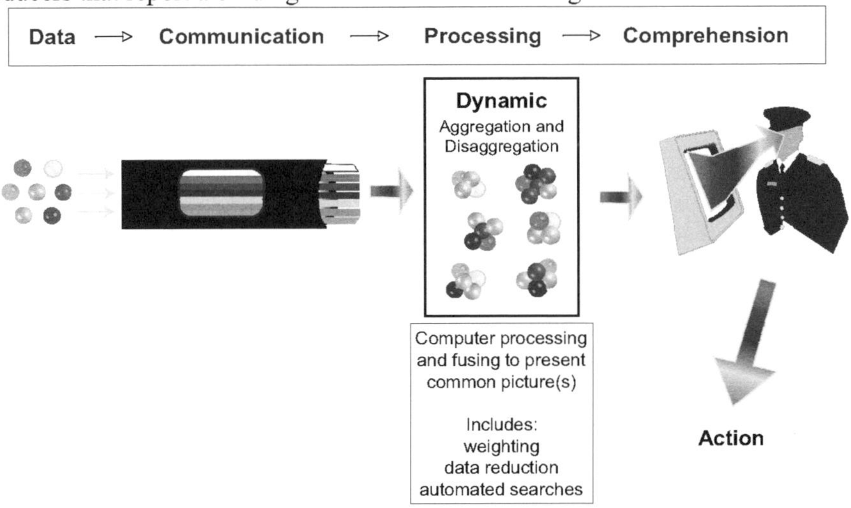

Figure 1. Information and decision making based on data processed through autonomous means using data fusion, which includes weighting, data reduction, and automated searches, to generate a readily understood iconographic format for final decision making and action.

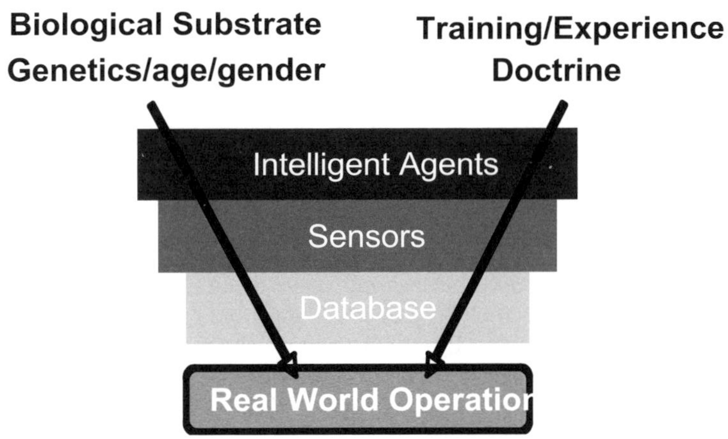

Figure 2. Databases of genomic and proteomic markers, host susceptibility, and baseline levels of biologic agents in the environment are compared to sensor acquired data, and interpreted by intelligent agents. Real world operations then act on the acquired intelligence by characterizing the biological substrate and utilizing training, experience, and doctrine to respond appropriately.

platform, and a transmitter reporting the results to some central data collector. The current availability of very high affinity immune based binders (Kd=10^{-13}) for infectious agents and of genomic probes will facilitate the development of effective genomic and proteomic sensor platforms. While extensive deployment of the sensor platforms in public arenas would provide useful data about dissemination of threat agents in specific communities, the inability to track individuals infected with the agent would diminish the capability of mitigating spread of disease. However, the distribution of large numbers of cell phones containing chips that would recognize threat agents in the sputum of users and send secure electronic transmission of positive responses detected by the device to a central autonomous data bank for comparison with archival data sets would permit tracking of infected individuals. This would facilitate management of disease in the individual and community as a whole. Under ideal conditions, the probes used in the sensor platforms would be designed to bind to emergent diseases on a worldwide basis (e.g. anthrax, SARS, West Nile Fever, Congo Crimean Hemorrhagic Fever, and smallpox), as well as on a local basis.

Secure telecommunications assets can link sensors at multiple locales to a central data processing center to validate any positive responses received from a single sensor; such validation would alert first responders and local health authorities to the potential threat. Archival health data sets and electronic medical records will help in the initial differential diagnosis process. For instance, detection of influenza virus by a sensor at a seasonal time when influenza usually emerges signals the probability that a new outbreak of infectious disease is in progress

The justification for implementing such a high cost sensor and data acquisition system is the conferred ability to respond swiftly enough to mitigate or deter the threat. Biological threats are a low probability, high-risk event; therefore, most of the sensor based events determined to be positive responses are likely to be false positives. This is particularly true for threats involving intentional release of biological agents, since there were less than ten such events over the past two decades. The benefit of widely distributing biological and chemical agent sensors is to detect emergent disease, accidental release of toxic chemicals, and potential terrorist actions, with a sufficiently high reliability to minimize the likelihood of false positives. Early warning systems allow pre-positioning of healthcare providers, pharmaceuticals, and materials for containing the spread of disease. Additionally, home care workers charged with the care of the aged and infirm can transmit information regarding infectious disease and receive treatment instruction via telecommunications systems (telemedicine). Improved community health also results from early warning systems, since proper treatment and precautionary measures can arrest the spread of contagions.

Each of these goals may be accomplished using sensor platforms deployed on public lands and properties including buses, planes, rail systems, and public toilets. The advantage of adding personal biosensors to the repertoire of biosensors for integrated probability assessment is the ability to track individual carriers of pathogens. These pathogen hosts may not express symptoms of disease, but may be able to spread the disease. By tracking the individual, epidemiologists can better understand the way in which disease spreads; therefore, they can obtain a better understanding of how to limit or respond to the threat. While sensors at bus terminals or airports could determine the presence of a pathogen without knowing who specifically is carrying that pathogen, there is no way to determine where it is going. Thus, there is no way to track how that pathogen affects public health.

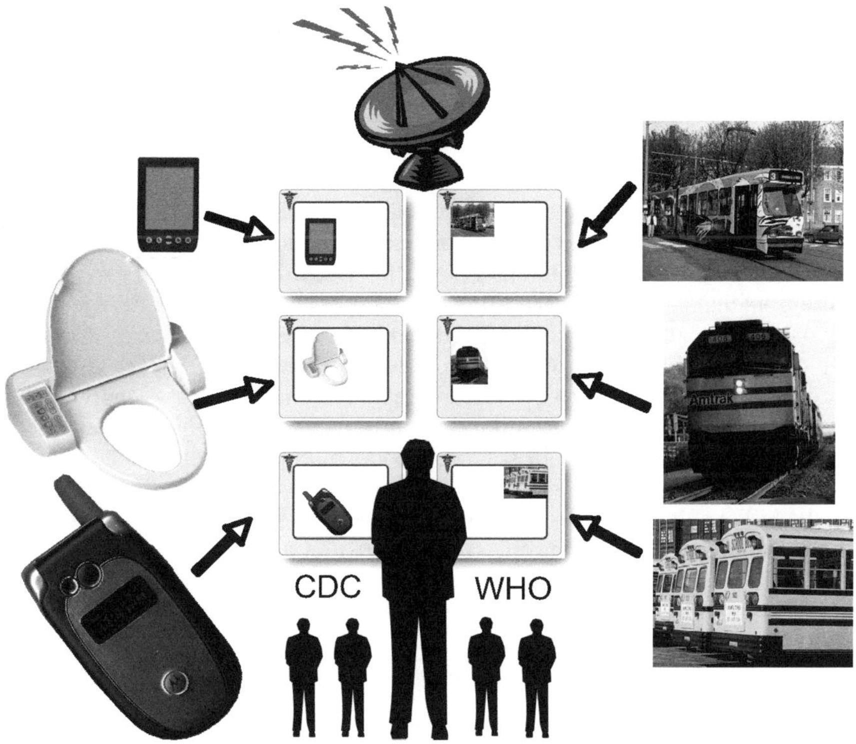

Figure 3. Data collected from a variety of personal and public biosensors are integrated and transmitted to a centralized health alert network.

2. Societal Impact

Correlates of probabilistic assessment include a large database of normal baseline conditions, pervasive and intrusive sensors, data acquisition, fusion, and interpretation. The combined capability of all these tools is at the heart of differential medical diagnostics. However, such pervasive use of sensors raises issues of privacy. In the normal transactions of any society, there exists a trade-off between personal freedom, privacy, and public safety. The challenge is to develop a coherent balance between ensuring public safety in a threat environment and respecting the potential loss of individual privacy.

The development of a compact chip able to bind selectively to threat agents and transmit that information electronically to an organization such as the CDC would have profound effects on improved healthcare delivery. The presence of such a chip in the home of isolated individuals would enable home healthcare for the infirm, measure healthcare delivery over distances, and detect the spread of infectious agents on continental dimensions. The societal issues raised by the deployment of these biosensors include potential loss of privacy of individuals resulting from the ability to pinpoint individuals with increased susceptibility or that have perceived socially objectionable diseases, such as AIDS or other STDs. The security vulnerabilities of such sensors increase the possibility of divulging personal medical information and could potentially expose individuals to increases in insurance rates or to loss of or compromised individual employment opportunities.

3. International Issues

With so many social concerns and so many potential benefits, social acceptance of distributed biosensors requires new policy and discussion at a level of the body politic. The concerns result from the potential loss of privacy due to the extensive deployment of biosensors that can identify specific vulnerability changes in identifiable individuals. Many nations are also concerned that information awareness increases threat potential by providing information on how to spoof the biosensors, which raises issues of public safety (i.e. mass panic, overrun hospitals) and economic security. Individual nations may perceive that wide distribution of information concerning emergent disease increases the risk of economic losses, such as the billions of dollars lost by nations that have experienced outbreaks of SARS or the negative effect of influenza on tourism. Also, the billions of dollars lost to Canada following detection of BSE in several cattle and to European nations during the outbreak of foot and mouth disease several years ago exacerbate this concern. However, many governments, consumers, and tourists will want to know the potential risks they are facing from emergent disease. Thus, biosensors must have a high degree of accuracy and very low false positive rates in order to convince governments to deploy them.

In the proposed dynamic assessment paradigm, emergence of infectious disease is a function of the level of the threat agent in an environment, the effective dissemination of the agent in aerosols or food, the environmental support of threat conditions, and the preparedness of a community to limit propagation of disease in that community. This probabilistic assessment assumes that there will be false positives and false negatives but will allow an information integrator to recognize a change in threat with a given probability. The larger the database and the more pervasive and intrusive the sensors and data acquisition system, the higher the probability is that a change in state of threat will be recognized.

4. Management of Concerns

If we are to develop a solution to these issues, several national and international protocols will need to be established. Countries employing public and private biosensors will need to develop national regulations regarding privacy issues. The Health Insurance Portability and Accountability Act (HIPAA) regulations developed in the US over the past ten years may serve as a model in this regard. Probabilistic based international standards regarding acceptable threat levels of biological agents in food and selected environments, such as hospitals, public transportation, and indoor/outdoor public areas, will need to be established. Additionally, the public trust factor of biosensor systems has not yet been evaluated. For all of these management strategies, trust must be established between a government and its population and between nation states engaged in large-scale economic transactions. Societal acceptance of a biosensor network with high costs and potential privacy issues will depend upon system reliability and transparency.

Conclusion

Emergent disease is a very low probability, high consequence event. Therefore, most of the positive indicators obtained from an early alert system will not be indicative of increased threat conditions. Effective measures of data collection require the distribution of sensors in both public places and on individuals. The mass deployment of sensors that can detect changes in levels of biological agents and of host susceptibility to these agents holds the possibility of early detection of increased threat, thus enabling initiation of effective countermeasures. At the same time, the distribution of the sensors can be perceived as potentially diminishing the assumed rights of individuals to privacy. This perception will probably be of greatest concern when changes in host susceptibility can be linked in a database to specific individuals (as with personal sensor platforms in cell phones). The data collected may influence insurance rates of individuals, employment opportunities, and negatively affect tourism and export on a national level. Developing a coherent policy that balances the right of an individual to maintain privacy with the necessity of promoting human health within and across government boundaries requires careful consideration of the implications of all possible options. The solutions developed will probably vary as a function of national culture, industrial development, and societal norms. Nonetheless, since biological agents do not recognize borders, the harmonization of policy regarding these threats and countermeasures should be initiated on a trans-national basis. This may involve a separate set of national policies that will complement agreed upon international policies for rapidly containing threats.

Acknowledgements

The authors would like to thank Dr. Rebecca Steinberg and Ms. Terisha Thomas for excellent support in editing and revising the manuscript. This work was supported under grant W911NF-07-2-0023.

L.M. Lechuga et al. (Eds.)
IOS Press, 2008

Portable nanobiosensor platforms for ultrasensitive multidetection of biological warfare agents in real time

Laura.G. CARRASCOSA, Elba MAURIZ, José SANCHEZ DEL RIO, Miguel MORENO, Kirill ZINOVIEV, Ana CALLE, Carlos DOMINGUEZ and Laura M. LECHUGA[1]
Microelectronics National Center (CNM). CSIC, Madrid, Spain

Abstract. For the development of field sensor systems with enough sensitivity and selectivity we develop photonic biosensors based on evanescent wave detection. Two technologies have been implemented in parallel: a plasmonic sensor for multianalyte real-time evaluation and an integrated optical nanosensor fabricated with silicon microelectronics technology. Both devices can be use as portable analyzers and have demonstrated an excellent performance in the immunological determination of chemical pollutants and in the detection of single mutations in DNA strands.

Keywords. Optical sensor, surface plasmon resonance, integrated optics, inhibition immunoassays, DNA hybridization.

1. Introduction

The progressive demand for the rapid and precise detection of any type of biological and chemical warfare agents (BCW) in air and water samples has accelerated the development of a large variety of biosensors. For this type of application it is desirable to have a selective biosensor of high sensitivity, fast response and able to perform real-time measurements. These requirements can be achieved mainly with optical sensors, due to the inherent nature of optical measurements that endow a great number of different techniques as emission, absorption, fluorescence, refractometry or polarimetry. Among them, photonic biosensors based on evanescent wave detection have demonstrated its outstanding properties as an extreme high sensitivity method for the direct recognition of substances in real time and in label-free schemes [1].

In the evanescent wave detection (see Figure 1), a receptor layer is immobilized onto the waveguide; the exposure to the partner analyte molecules produces a biochemical interaction, which induces a change in its optical properties. This change is detected by the evanescent wave. The extent of the optical change will depend on the concentration of the analyte and on the affinity constant of the interaction, obtaining a quantitative signal of the interaction. The evanescent wave decays exponentially as it penetrates the outer medium and, therefore, only detects changes taking place on the

[1] Corresponding author: L.M. Lechuga, IMM-CNM, CSIC. Isaac Newton, 8. 28760 Tres Cantos, Madrid, Spain. E-mail: laura@imm.cnm.csic.es

surface of the waveguide. For that reason it is not necessary to carry out a prior separation of non-specific components (as in conventional analysis) because any change in the bulk solution will hardly affect the sensor response. In this way, evanescent wave sensors are selective and sensitive devices for the detection of very low levels of chemicals and biological substances and for the measurement of molecular interactions in-situ and in real time [2].

The advantages of the optical sensing are significantly improved when this approach is used within an integrated optics context [3]. The technology of integrated optics allows the integration of passive and active optical components (including fibers, emitters, detectors, waveguides and related subsystems) onto the same substrate, allowing the flexible development of miniaturized compact sensing devices, with the additional possibility to fabricate multiple sensors on a single chip.

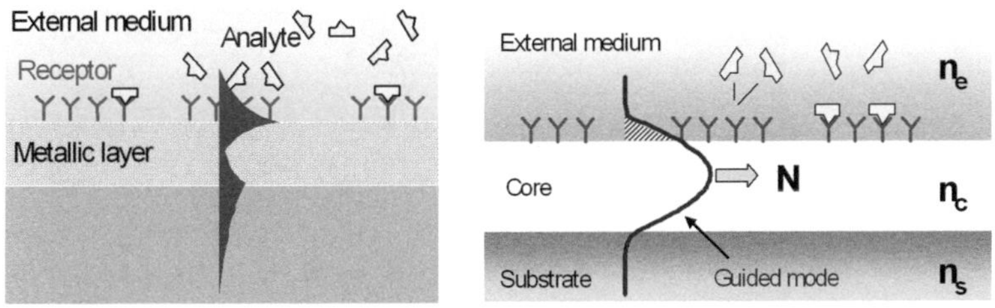

Figure 1. Evanescent field sensors and the detection of biomolecules with (a) SPR sensor and (b) planar waveguides sensors.

The biological receptor for the detection of BCW agents could be or DNA strands that can bind to a specific pathogen present in the environment, either antibodies that can recognize specific sites on bacteria or bind to surface proteins. The optical biosensors to be developed today and in the near future should be able to work with both types of receptors as well as to trigger a signal after their specific detection.

In order to analyze the main components and functionality of a multibiosensor technological platform that could be used as an early warning system for biological and/or chemical warfare, we will show the applicability of two different optical biosensor technologies:

- a platform based on a portable Surface Plasmon Resonance Sensor (SPR)
- a platform based on integrated nanointerferometer devices incorporating the microfluidics.

2. Surface Plasmon Resonance Biosensor

2.1. Sensor set-up

One of the best known and more developed optical biosensors is the Surface Plasmon Resonance (SPR) sensor [4], because of its sensibility and simplicity. Surface plasmons are elementary excitations, which result from a collective oscillation of the free electron plasma at a metal-dielectric film interface. In a SPR sensor a thin metal film (usually Au) is evaporated on the dielectric material surface. The sensing mechanism is based on variations of the refractive index of the medium adjacent to the metal sensor surface during the interaction of an analyte to its corresponding receptor, previously immobilized at the sensor surface in the region of the evanescent field. The recognition of the complementary molecule by the receptor causes a change in the refractive index and the SPR sensor monitors that change.

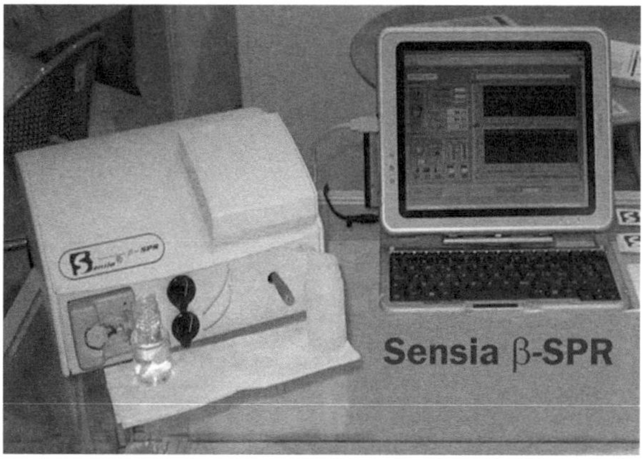

Figure 2. Portable SPR platform, which includes the sensor module, the microfluidics, and the electronics and data handling.

We have developed a portable SPR sensor prototype (see Fig. 2). The configuration of this SPR system has been reported in previous papers [5]. The resonant condition is achieved in the Kretschmann configuration by means of a prism coupler structure. A laser beam emitting at 670 nm from a 3 mW laser diode source is divided into two equal beams to allow two-channeled simultaneous measurements onto the gold-coated film used as sensing surface. For monitoring binding events in real time, SPR measurements are performed at a fixed angle of incidence.

2.2. Immunosensing of chemical pollutants

As a proof of its utility for the detection of pathogens, the device has been applied as a highly sensitive field analytical method for environmental monitoring of several pesticides including the chlorinated compound DDT, the neurotoxins of carbamate type (carbaryl), and organophosphorus compounds (chlorpyrifos) [6]. Organophosphate compounds are well-known irreversible inhibitors of acetylcholinesterase at cholinergic synapses that cause the disruption of nerve signal transduction. They have been considered as powerful neurotoxins with a threatening potential as chemical warfare agents.

For the detection of these chemical substances an SPR immunoassay has been implemented using specific monoclonal antibodies. But most of the substances of interest at the environmental, clinical and food industries are of small molecular weight (MW< 5000 Da) and are difficult to be detected by using a direct or sandwich immunoassays but they can be detected by using an inhibition assay. In an inhibition assay (see Fig. 3 for details), a low-molecular weight analyte (antigen) is immobilized on the sensor surface. Sometimes this small molecule is previously attached to a carrier protein (as BSA or OVA). After, the sample is incubated with the specific antibody for a short time (a few minutes), allowing the analyte to be recognized. The concentration of the antibody must be kept constant in all the experiments. Finally, the incubated solution is flowing over the sensor surface, in such a way that the remaining free antibody can recognizes the immobilized antigens. Depending of the analyte concentration in the incubated solution, there will be more or less free antibody available. The signal in the sensor is, therefore, inversely proportional to the concentration of the analyte in solution.

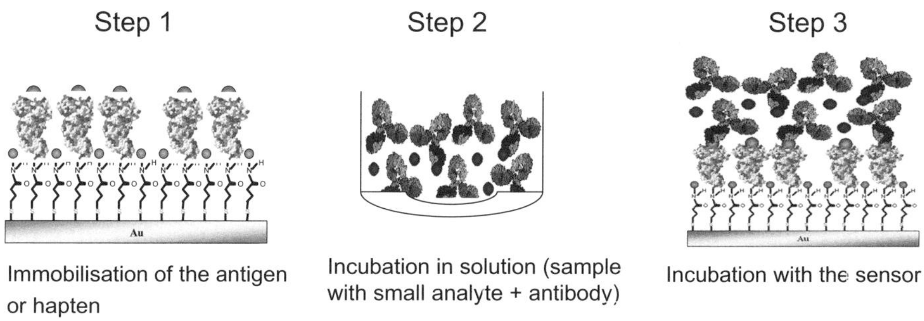

Figure 3. Steps in an inhibition immunoassay

The immobilization of the bioreceptors involves the formation of self-assembled monolayers (SAMs). First, amine groups of the recognition element were coupled to the carboxylic terminal groups of the alkanethiol monolayer through a carbodiimide linkage. The formation of carboxylic terminated SAMs involved the use of mercaptoundecaonoic acid at 0.05 mM in ethanol that was flowed over the gold sensing layer. The activation of the alkanethiol carboxylic groups to a stable intermediate (N-hydroxysuccinimide ester) was accomplished using a mixed solution of EDC/NHS (0.2/0.05M in distilled water). In this state, the modified sensor surface is easily available for the amine groups of the BSA-hapten conjugate used as recognition element throughout two consecutive injections. Undesirable effects due to non-specific binding and formation of conjugate multilayers were counterbalanced by using a solution of ethanolamine 1M, pH 8.5 as blocking agent.

Applying a binding inhibition immunoassay, for DDT, the dynamic range of the sensor was 0.06-14 μg/l, with a limit of detection of 0.02 μg/l. For carbaryl, the dynamic range of the sensor was 1.6-14.7 μg/l, with a limit of detection of 1.36 μg/l. Likewise, the immunoassay for chlorpyrifos determination, afforded a high sensitivity working in the 0.2-28 μg/l range with a detection limit of 0.05 μg/l. Figure 4 shows the calibration curves obtained from the three pesticides.

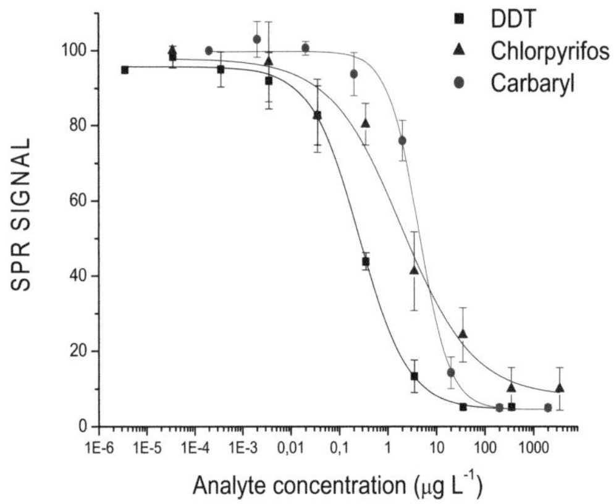

Figure 4. Calibration curves of three pesticides as determined by inhibition SPR immunoassay

2.3. Analysis of real water samples and multianalyte format

The influence of different water types (groundwater, tap and river water) on the immunosensor response was evaluated. The ionic strength of water samples did not need to be adjusted and spiked water samples were prepared like the standard samples. As an example, Figure 5 shows the chlorpyrifos standard curves obtained regardless of the nature of the water samples. Very similar LOD values were obtained for groundwater, river and tap water, respectively. Consequently, pesticide determination seemed to be directly applicable to environmental waters without any sample pretreatment as matrix effects were not observed. Similar results were obtained for DDT and carbaryl analysis in real samples [7].

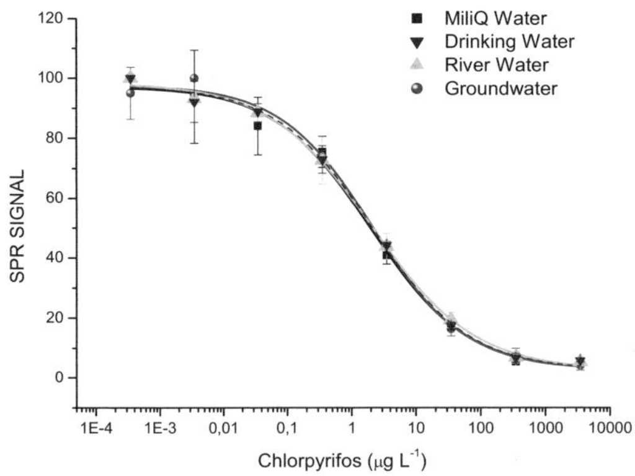

Figure 5. Comparison of the standard calibration curves for chlorpyrifos in different water types: groundwater, river and tap water.

Although the SPR system has only two channels, we have implemented a new procedure for simultaneous multi-analyte determination. The approach is based on the multiple and combined spatial controlled immobilization of up to three analyte recognition elements on the same sensing surface. In this format, analyte determinations were carried out using sequential modes by flowing chlorpyrifos, carbaryl or DDT samples over the same channel. Due to the difference in the hapten to protein molar ratio the optimum concentration for each analyte conjugate to be immobilized should be previously determined. The optimal combination of multiconjugate concentrations were obtained for 15, 10 and 5 μg.mL^{-1} dilutions of DDT, carbaryl and chlorpyrifos conjugates, respectively.

The calibration standard curves are obtained in an ordered sequence in which serial dilutions of one of the target analytes will be followed by that corresponding to the other two pesticides. The stability and reproducibility of the immunosensor response was not affected by the order of antibody injections. An assay cycle including each antibody interaction and their corresponding regeneration cycles takes 60 min
Limits of detection of 18 ng.L^{-1} and IC50 value of 0.44 μg L^{-1} were attained for DDT detection in this multianalyte format. Chlorpyrifos inhibition binding curves (see Figure 6) also reached an extremely similar linearity, and low detection limits (52 ng L^{-1}) and IC50 (1.76 μg L^{-1}) values. Finally, the assay sensitivity obtained for carbaryl shows a significant improvement of the detection limit when comparing the multi-analyte format (0.05 μg L^{-1}) to the single analyte one (1.41 μg L^{-1}).

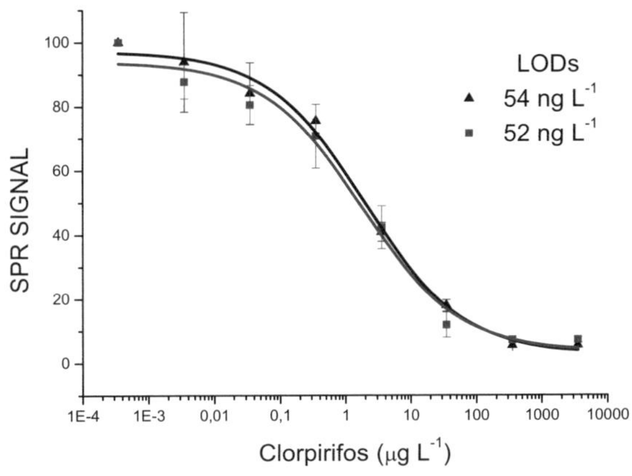

Figure 6. Comparison of the standard calibration curves for chlorpyrifos for single (*triangles*) and multianalyte formats (*squares*). Measurements were done in triplicate.

The analysis time for all three pesticides varied from 40 to 60 min depending on the number of regeneration cycles. The assay reproducibility was proved through the repeated use of the same sensor surface for over more than 200 assay cycles, whereas the absence of biosensor response to non-related analytes showed the specificity and reliability of the analysis. The influence of cross-reactants was also evaluated by flowing samples containing all three analytes. The immunosensor response deviation from the expected signal in the absence of non-specifics analytes was negligible, showing the specificity of the multi-surface to recognize only the target analyte.

The SPR multi-surface approach has shown to be effective for determining the target analytes in a highly sensitive, simple and rapid manner, without the need of labeling biomolecules or spatially resolved surface modification. These are inherent advantages over chromatographic and fluorescence-based multi-biosensors. In addition, this method could be extended to the detection of more than three analytes by simply modifying the sensing surface with a higher number of compounds. The multi-biosensing system offers a versatile, regenerable, robust, fast and cost-effective field analytical method for the monitoring of chemical and biological warfare compounds if the corresponding bioreceptor is available.

2.4. Evaluation in human samples

The SPR sensor can also be used as an on-line immunoanalytical method for the evaluation of pesticide metabolites which can be present in the human body [8]. We have detected the metabolite from chlorpyrifos pesticide (3,5,6-trichloro-2-Pyridinol (TCP)) from its primary source of elimination (urine). This metabolite can be detected in urine samples several days after exposure to the pesticide.

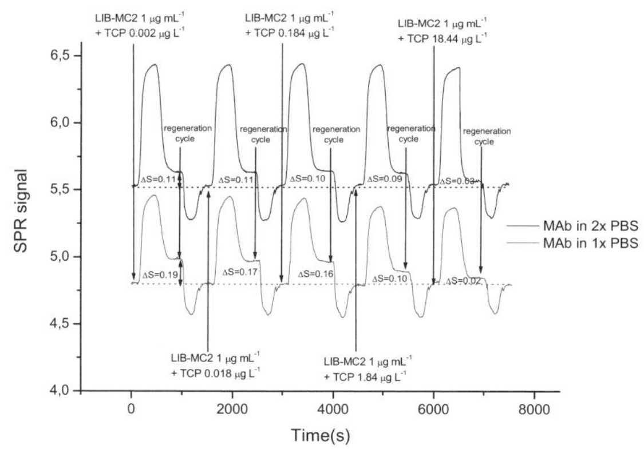

Figure 7. SPR detection of TCP in human urine. A regeneration cycle is applied after each determination. A limit of detection of 0.1 $\mu g.L^{-1}$ was obtained.

Serial dilutions of TCP (0.002-184 $\mu g.L^{-1}$), were prepared in PBST 1x, distilled water or human urine from a stock solution in DMF of 0.2 g L^{-1}, and mixed with a fixed concentration of the corresponding monoclonal antibody (1:1; v/v). Before injection, final dilutions were incubated for 10 minutes at room temperature. SPR response to the binding between antibody in solution and the immobilized hapten is monitored in real-time. To achieve a reusable surface, antibody-hapten associations were disrupted by regenerating the biosensing layer with 0.1 M HCl. Standard calibration curves were obtained by averaging three individual curves. Verification of the selectivity of the assay was confirmed by measuring the effect of non-specific antibodies to the immobilized hapten. Human urine samples were collected from a healthy adult volunteer over 2 days. Samples did not need to be reconstituted by diluting urine concentration. This solution was fortified with TCP at different concentrations to perform binding inhibition tests in urine. From a 0.2 g L^{-1} TCP stock solution, urine samples under study were spiked with TCP at 0.002, 0.018, 0.184, 1.84, 18.44 and 184.43 $\mu g.L^{-1}$.

Binding inhibition tests were performed then in the untreated urine samples and compared to those obtained in distilled water and PBS used as control (see Figure 7). In all cases, similar detection limits, in the micrograms per litre level (0.1-0.24 $\mu g.L^{-1}$), were attained for TCP assays independently of the dilution buffer. The reproducibility of measurements was studied throughout more than 130 regeneration cycles, allowing the repeated use of the same immunosensor surface without significant variation of the SPR signal. All measurements were developed in real-time in only 10 minutes.

The immunoassay format has shown to be a sensitive detection of TCP in human urine without the need of previous clean-up and preparation of samples. Our SPR direct detection approach to TCP quantification in urinary samples could be easily transferred to other dangerous relevant substances, drug residues, veterinary products or exposure markers susceptible to be found in body fluids (blood, urine). The simplicity, low time of response and real-time determination are inherent advantages over chromatographic and immunochemical-based methods.

3. Integrated interferometers

3.1. Sensor development

Evanescent wave photonic biosensor devices based on standard microelectronics and related micro/nanotechnologies are providing an integrated technological solution for achieving high sensitive arrays of biosensing devices. But problems of stability, sensitivity and size have prevented the general use of integrated optical biosensors for real field applications.

In order to solve the above drawbacks, we are developing ultrasensitive and miniaturized photonic silicon sensors that may be integrated into a “lab-on-a-chip” microsystem platform. As sensors we use integrated Mach-Zehnder interferometer based on TIR waveguides ($Si/SiO_2/Si_3N_4$) of micro/nanodimensions as it is shown on Figure 8.

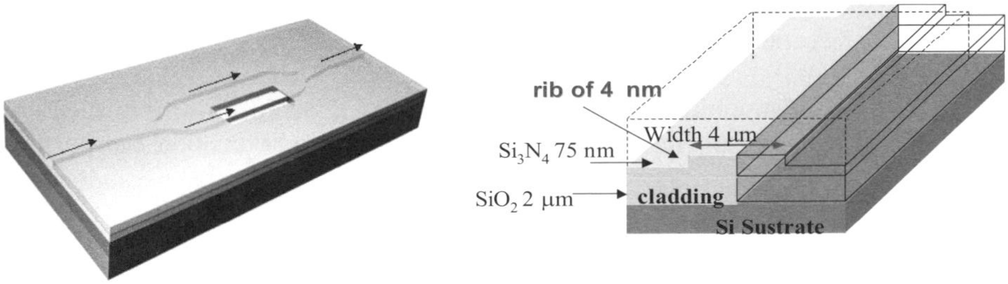

Figure 8. Mach- Zehnder interferometer configuration (*left*). Cross-section scheme of the TIR waveguides employed in the integrated MZI device (*right*)

In a Mach-Zehnder interferometer (MZI) device the light from a laser beam is divided into two identical beams that travel the MZI arms (sensor and reference areas) and are recombined again into a single mode channel waveguide giving a signal which is dependent on the phase difference between the sensing and the reference branches. Any change in the sensor area (in the region of the evanescent field) produces a phase difference (and therein a change of the effective refractive index of the waveguide) between the reference and the sensor beam and then, in the intensity of the out coupled light. For the evaluation of specific biosensing interactions, the receptor is covalently attached to the sensor arm surface, while the complementary molecule binds to the receptor from free solution. The recognition of the complementary molecule by the receptor causes a change in the refractive index and the sensor monitors that change.

For biosensing applications the waveguides of the MZI device must be designed to work in the single mode regime and to have a very high surface sensitivity of the sensor arm towards the biochemical interactions. The highest surface sensitivity is obtained in waveguides with a high contrast of refractive index between the core and the substrate. For that reason, we have chosen Si_3N_4 core layers ($n_c = 2.00$) over a SiO_2 substrate ($n_s = 1.46$). In this waveguide configuration (and for λ in the visible range), the single mode behavior is obtained for core thickness below 300 nm, and rib depths below 5 nm, when the rib width is 4 μm. Our modeling shows that the maximum surface sensitivity is obtained for core thicknesses about 150 nm and 75 nm for the TM and TE polarizations, respectively [9].

For evaluating the sensor sensitivity a calibrating curve was recorded using solutions with different refractive indexes. The lowest detection limit which can be detected with this device in the variation of the refractive index was found to be (for TM polarization) $\Delta n_{0,\,min} = 1 \cdot 10^{-7}$ ($N_{eff,\,min} = 6.4 \cdot 10^{-8}$) which means the possibility to detect 60 fg/mm^2 in a direct way.

3.2. DNA testing

As a proof of the utility of MZI technology towards the detection of ultra-sensitive biomolecular interactions, we have applied the sensor for the direct detection of DNA hybridization and for the detection of single nucleotide polymorphisms at the BRCA-1 gene, involved in breast cancer development, without target labeling.

The oligonucleotide probe is immobilized by covalent attachment to the sensor silicon nitride surface through silanization procedures. A silane (3-mercaptopropyltrimethoxysilane) with a thiol group at the free end was employed for the chemical modification of the surface. The thiol-derivatized oligonucleotides (28 mer) used as receptors can bind to the silanized Si_3N_4 surface through a disulphide bond.

After DNA immobilization, complementary oligonucleotides (58 mer) were flowing in the sensor for hybridization experiments. The hybridization was performed for different DNA target concentrations from 1 pM to 1 μM. Non-complementary oligonucleotides did not show any significant signal. Regeneration after each hybridization was achieved flowing DI water and HCl 3.2 mM. The calibration curve can be observed in Figure 9. 10 pM of complementary non–labeled DNA in buffer solution was the lowest hybridization limit achieved. Additionally, we have detected the hybridization of 100 nM DNA target with two mismatching bases corresponding to a mutation of the BRCA-1 gene.

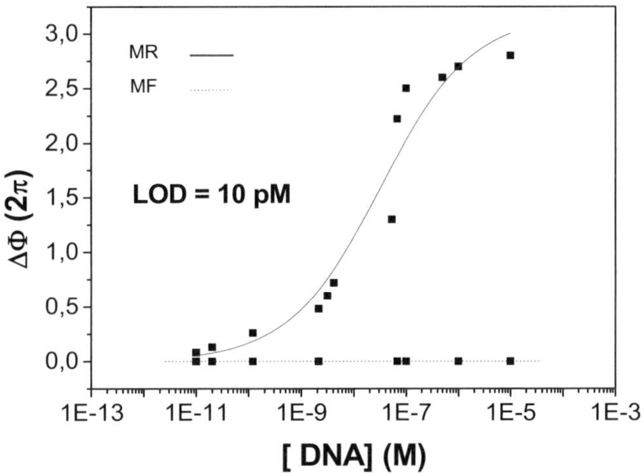

Figure 9. Calibration curve of the DNA hybridization evaluated in real-time by the MZI device.

3.3. *Lab-on-a-chip integration*

Devices for on-site analysis or point-of-care operations for biological and chemical warfare detection are geared for portability, ease of use and low cost. In this sense, integrated optical devices offer a compact structure and could allow for fabricating optical sensor arrays on a single substrate for simultaneous detection of multiple analytes. The main advantage of the development of Mach-Zehnder devices is the possibility to develop a complete "lab-on-a-chip" by optoelectronic integration of the light source, photodetectors and sensor waveguides on a single semiconductor package together with the flow system and the CMOS electronics.

For the development of a complete lab-on-a-chip microsystem device based on integrated MZI, several units must be incorporated on the same platform: (i) the micro/nanodevices, (ii) the flow cells and the flow delivery system, (iii) a phase modulation system to convert the periodic interferometric signals in direct phase measurements, (iv) integration of the light sources and the photodetectors (v) CMOS processing electronics. For achieving this goal, our first step has been the development of a novel low temperature (100 ºC) CMOS compatible microfluidic technology to create 3D embedded interconnected microfluidic channels between different substrates [9] using SU-8 as structural material. The microfluidic channels have a height from 40 to 60 μm and a width between 100 to 250 μm. First experiments with the integrated microfluidic-MZI sensor have shown a sensitivity close to that obtained without the integration of the microfluidic [10] and this opens the way for further device integration in a truly hand-held microsystem.

4. Conclusions

For the rapid detection and identification of biological and chemical warfare agents, reliable multi-biosensor systems allowing label-free and real time measurement of simultaneous interactions must be developed. We have discussed the advances in the development of different optical biosensor platforms: a portable surface plasmon resonance Sensor (currently in commercialization) and an integrated Mach-Zehnder interferometer device made on Si technology. The feasibility of the different biosensor platforms have been proved by the immunological recognition of several pesticides including the chlorinated compound DDT, and the neurotoxins of carbamate type (carbaryl) and organophosphorus type (chlorpyrifos which resembles SOMAN chemical warfare agent). These results open the way for further development of portable and multianalyte platform for the detection of several biological and chemical warfare agents in-situ and in real-time.

5. Acknowledgements

Authors would like to thank the financial support of the Spanish Ministry of Education and Science, M. Botín Foundation and the Spanish Research National Council (CSIC). Authors thanks also to Dr. Angel Montoya (Ci2b, Univ. Politécnica de Valencia) for supplying the immunoreagents.

6. References

[1] Lechuga, L.M., (2005) Optical Biosensors, in *Biosensors and Modern Biospecific Analytical Techniques*, Comprehensive Analytical Chemistry Series Vol. XLIV, L. Gorton, Ed., Elsevier Science BV, Amsterdam, Netherlands

[2] F.S. Ligler and C. Rowe Taitt (Eds.), Optical Biosensors: Present and future. Elsevier, Amsterdam, 2002.

[3] C. Domínguez, J.A. Rodríguez and L.M. Lechuga. In: S. Alegret (Ed.), Integrated Analytical Systems. Elsevier, Amsterdam, 2003

[4] L.M.Lechuga, E. Mauriz, B. Sepúlveda, J. Sánchez del Río, A. Calle, G. Armelles and C. Domínguez. *Optical Biosensor devices as early detectors of Biological and Chemical Warfare agents In* Frontiers in Planar Lightwave Circuit Technology (design, simulation and fabrication). NATO Science Series. Mathematics, Physics and Chemistry, 216 (2006). Ed. S. Janz. Springer. Dordrecht, The Netherlands

[5] E. Mauriz, A. Calle, A. Abad, A. Montoya, A. D. Barceló and L.M. Lechuga, Determination of carbaryl in natural water samples by a Surface Plasmon Resonance flow-through immunosensos. Biosensors& Bioelectronics 21(11) (2006)

[6] E. Mauriz, A. Calle, A. Montoya and L.M. Lechuga. Determination of environmental organic pollutants with a portable optical immunosensor. Talanta 69 (2) (2006)

[7] E. Mauriz, A.Calle, J.J. Manclús, A. Montoya, A. M. Escuela, J.R. Sendra and L.M. Lechuga. Single and Multi-analyte Surface Plasmon Resonance assays for simultaneous detection of cholinesterase inhibiting pesticides. Sensors and Actuators B. 118 (2006)

[8] E. Mauriz, A.Calle, J.J. Manclús, A. Montoya, and L.M. Lechuga. On-line determination of 3,5,6-trichloro-2-Pyridinol in human urine samples by Surface Plasmon Resonance immunosensing. Anal. Bioanal. Chem. 387, 2757-2765 (2007)

[9] B. Sepúlveda, J. Sánchez del Río, M. Moreno, F.J. Blanco, C. Domínguez and L. M. Lechuga. Optical biosensor microsystems based on the integration of highly sensitive Mach-Zehnder interferometer devices. J. of Optics A: Pure and Applied Optics, 8 (2006)

[10] F. J. Blanco, J Berganzo, M. Agirregabiria, K. Mayora, J. Elizalde, A. Calle, C. Dominguez and L.M. Lechuga. Microfluidic-Optical Integrated CMOS compatible devices for label-free biochemical sensing. J. of Micromechanics and Microengineering, 16 (2006)

L.M. Lechuga et al. (Eds.)
IOS Press, 2008

Development and Testing of the Portable Electrochemical Immunosensor System for Detection of Bioagents

Petr SKLÁDAL [a,b,1] , Jan PŘIBYL [b], Bohuslav ŠAFÁŘ [c]
[a] *Institute of Biochemistry and* [b] *National Centre of Biomolecular Research, Masaryk University, Brno, Czech Republic*
[c] *Military Technical Institute of Protection, Brno, Czech Republic*

Abstract. For the field detection of various microbial species, the portable immunosensor system was developed. The detector consists of a digital 4-channel potentiostat, flow-through system with 4 miniperistaltic pumps, microcontroller unit, rechargeable battery and flow-through cell with an exchangeable immunosensor. This compact system is controlled from external computers using either serial cable or Bluetooth wireless link. Software allows both manual operation and script-based automated measuring procedures. The experience from laboratory and field trial testing will be reported.

Keywords. Amperometric sensor, screen printed electrode, biosensor, flow through system, immunochemical detector

Introduction

The current requirements on detection of bioagents in different situations including military operations in field, civil rescue and security units, public buildings and homeland protection, stimulate development of portable, rapid and simple instrumentation based on the bioanalytical detection principles [i, ii]. For microbial agents, various types of immunochemical devices are preferred for the early response, good sensitivity and continuous monitoring capabilities. The detection occurs on the phenotype level, thus no extraction of the genetic material from the agent is required, which is the case for methods based on the polymerase chain reaction (PCR, [iii, iv]). Immunosensors applied for biodetection purposes operate mainly using various optical systems as transducers [v], surface plasmon resonance (SPR) [vi], e.g. Spreeta (www.sensata.com/products/sensors/spreeta.htm) and Biacore (www.biacore.com) representing the most popular system and a portable example. Alternative optical systems include fluorescence with evanescent wave excitation based on planar (Array Biosensor from Naval Research Laboratory, www.nrl.navy.mil [vii]) and cylindrical waveguides – the Raptor system from Research International (www.resrchintl.com/raptor-detection-system.html [viii]. A combination of both optical and electric principles is electrochemiluminescence which is successfully used

[1] Corresponding Author: Petr Skládal, Institute of Biochemistry, Masaryk University, Kotlářská 2, 611 37 Brno, Czech Republic; E-mail: skladal@chemi.muni.cz

in the M1 series biodetectors from BioVeris (www.bioveris.com/homelandsecurity/instruments/m1m.htm [ix]). Electrochemical sensors for bioagent detection represent a promising alternative approach [x,xi]; the electrochemical measuring system is affordable and easily miniaturized to a portable format. As the measuring element, the screen-printed sensors are widely applied [xii] due to easy and reproducible fabrication at both laboratory and mass production scales. This contribution describes the present state of our system ImmunoSMART, previously designed in collaboration with MultiLab, Brno [xiii] and recently completed in the Smart Ltd. company (www.smartbrno.cz).

Experimental

Measuring Instrument Construction

The ImmunoSMART detector (Figure 1) is a digitally controlled four-channel potentiostat (1.25 μA range, ±1 V excitation potential) including up to four miniature peristaltic pumps. The central microcontroller 89C52 (Atmel) ensures timing, internal data transfers with A/D (effectively >20 bit), D/A (16 bit) converters and digital I/O interfaces, communication with the external computer through serial link. The speed control of pumps is realized with two additional controllers 89C2051 (Atmel). Power for 8 to 24 hours of operation is provided by the internal rechargeable battery. The dimensions of the unit are 55 x 240 x 250 mm (height x width x depth) and it weights 1350 g.

Figure 1. Front view of the ImmunoSMART detector showing four embedded peristaltic pumps, control Ligth Emitting Diodes (LEDs) and connector for the measuring cell.

The rear panel (not shown) provides connector for the serial RS232C link, the connector is custom modified providing additional power (+5 V) for an optional external Bluetooth dongle. There is also another connector for external peripheral devices (4 TTL out and 2 TTL in signals, +5 V power), charging adapter connector, start button and power off switch.

Operating Software

For control of ImmunoSMART, the software ML-615x is used under the Windows operating systems, the standard (COMn) port is required, though virtual Bluetooth and

TCP/IP simulated serial ports can be used, too. The program is shown in Figure 2, the main task is setting of device options (working potentials, speed of pumps, connection of sensors, digital outputs) and continuous readout (data sampling at 100 ms), graphic display and storage of measured values of current in the four channels, plus temperature.

Alternatively, the operation can be fully automated using scripts edited from the program or using any text editor. The script commands include setting of potential SET_E1 to _E4, pump speed PUMP_1 to _4, START / STOP / PAUSE / CONT of data saving, WAIT period for timing of events, SENSOR ON / OFF / TEST, DIO setting, STANDBY and WAKEUP to save power when measurement is not required.

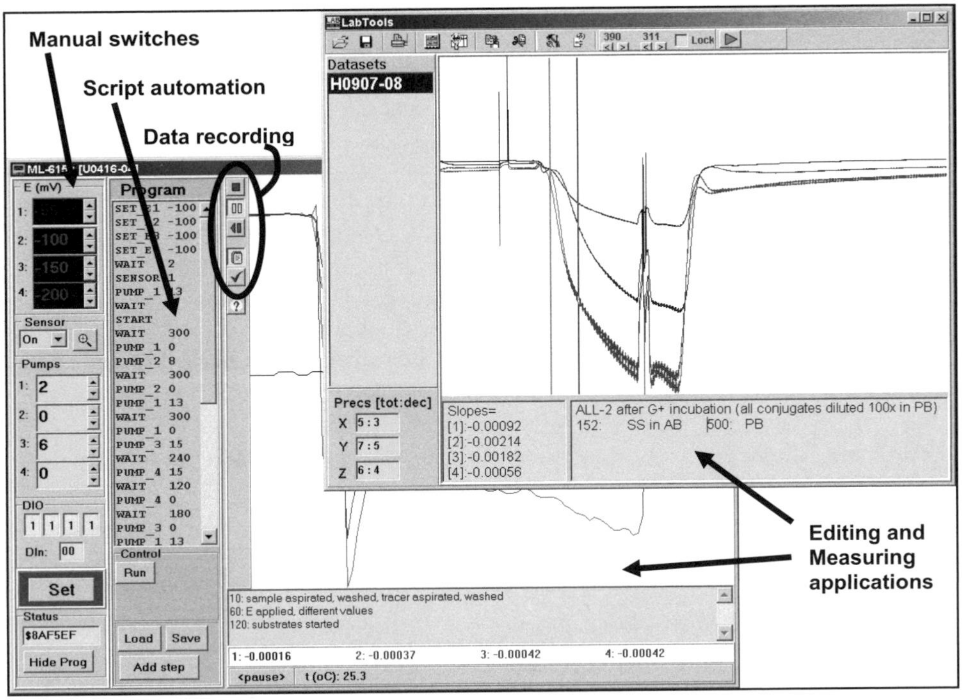

Figure 2. Software consists of the measurement control module (bottom left, ML-615x) and the dataset editing module (top right, LabTools, not required for measurement).

The once saved dataset (consists of data and accompanying comments) can be graphically edited using the Lab Tools program (Figure 2). A preliminary measurement control program compatible with the PocketPC (Windows Mobile) is available, too.

Immunosensors and Measuring Procedure

The construction of immunosensors based on the four-channel screen-printed sensors (supplied by BVT Technologies, www.bvt.cz) with gold electrodes was described previously [xiii], the measuring procedure is schematically shown in Figure 3.

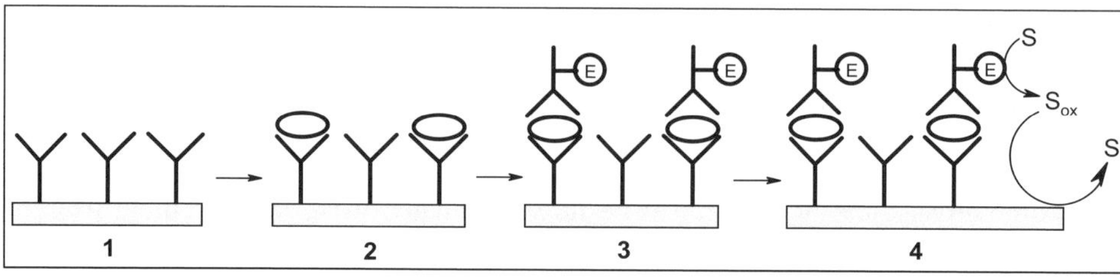

Figure 3. Scheme of the electrochemical immunoassay format. The gold surface with immobilized antibodies (1) is incubated with the sample and the target microorganism becomes captured (2), furthermore, sandwich with the secondary antibody labeled with peroxidase is formed (3) and finally the enzyme bound in the immunocomplex reacts with substrate (hydrogen peroxide and iodide) providing electrochemically active product – iodine, which is reduced at -100 mV (4).

Several types of immunosensors were developed; the scheme of immobilization of individual antibodies is presented in **Table 1**. The antibodies used against the target bioagent simulants were as follows: Biogenesis (www.biogenesis.co.uk) supplied anti *Erwinia herbicola* (EH) 278/427 and anti *Escherichia coli* (EC) 4329-4906; Abcam (www.abcam.com) supplied anti *Bacillus* (BA) BDI380 and anti *E. coli* ab20386; Tetracore (www.tetracore.com) supplied anti *E. herbicola* TC-7006 and anti ovalbumin (OVA) TC-7003. Furthermore, anti human serum albumin (HSA) AL-01 was obtained from Exbio Praha (www.exbio.com) and anti 2,4-D E2G2 was from Veterinary Research Institute (www.vri.cz), both these antibodies served as controls. Immobilization of antibodies was through the covalently linked protein A and antibody-peroxidase conjugates were produced as described previously [xiii].

Table 1. Immunosensors for testing of bioagents detection, antibodies on electrodes of the 4-channel sensors.

Chan.	**Immunosensor type**			
	EH	**OVA**	**BEC**	**ALL**
1	control (anti HSA)	control (anti HSA)	BA BDI380	EH 278/427
2	control 2 (anti-2,4-D)	control 2 (anti-2,4-D)	control (anti HSA)	control (anti HSA)
3	EH 278/427	OVA TC-7003	EC 4329-4906	BA BDI380
4	EH TC-7006	OVA TC-7003	EC ab20386	OVA TC-7003

Measurements were carried out either in the laboratory where the sample was simply aspirated to the measuring cell, or the sample was obtained as a flow of liquid collected by the cyclone device.

Results and Discussion

Instrument Testing

Originally, the target microorganism was *Francisella tularensis*, here, the measurements were carried out with safe microorganisms suitable for work in common laboratories and as simulants for field testing. In this way, the experiments need not to be carried out in the strictly controlled and certified biosafety level 3 microbiological facilities, which was rather impractical for the development phase. The common laboratory set-up of the detector with external flow-through cell containing the immunosensor is shown in Figure 4.

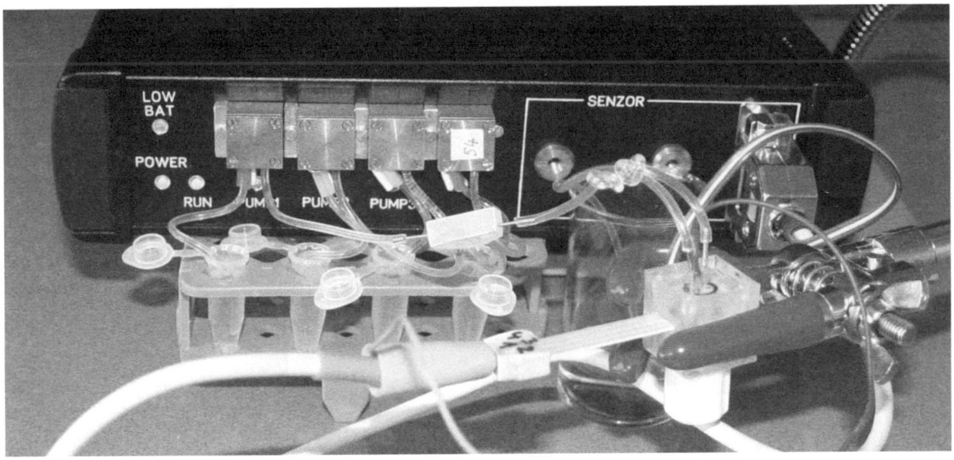

Figure 4. ImmunoSMART device used in laboratory conditions, the flow-through cell with inserted screen printed sensor is shown.

At present, the external parts of the measuring system are not finalized and the solution containers, flow-through cell and the sensor are loosely located in front of the detector. In the final version, this should be arranged as a compact component allowing a simple insertion of the immunosensor.

The immunosensor-based assay developed and optimized previously for *Francisella* detection [xiii] was extended to other microbial species, according to the requirements of the planned field trial. The idea was that by a simple exchange of the immobilized antibody, and using the appropriate tracer, a new immunosensor should be obtained. This approach was tested with two antibodies against *E. coli*, one antibody with a broader specificity against *Bacillus* species and a control anti albumin antibody, all immobilized in individual positions on the same strip. A positive response was obtained with highest values of current in both anti *E. coli*-specific channels; however, quite significant current values were measured also in the anti *Bacillus* and control channels (Figure 5).

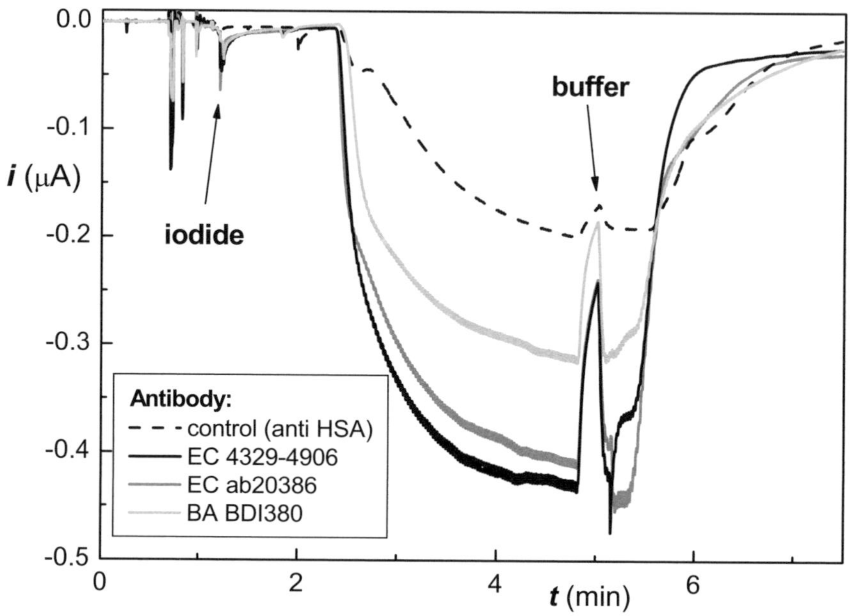

Figure 5. Examples of response traces from the immunosensor with immobilized anti *E. coli* (EC) and anti *Bacillus* (BA) antibodies, including anti albumin (HSA) as a control. Sample contained 10^3 CFU of *E. coli*, contact time 10 min, 5 min wash with 50 mM phosphate pH 7.0 containing 0.05% albumin. Followed a 5 min zone of the tracer – anti EC Ab-peroxidase conjugate (5 μg/ml) and buffer washing. The response was measured in 50 mM acetate pH 4.5 containing 1 mM H_2O_2, iodide (2 mM) was added as co-substrate and the cathodic current (-100 mV) was recorded.

This means that non-specific adsorption of tracer, microbe or of both components in the control channels could occur. Such problems should be addressed during the optimization period, which consequently can not be neglected.

Field Testing Experience

So far, the immunosensor detector was tested in laboratory using model buffer-based samples containing known amounts of microorganisms. For the field testing, only the type of the bioagent is known, but the concentration can be highly variable. As the bioagent field trials are not organized in our country because no suitable facilities exist here; we have therefore joined the trial organized by the NBC and FOI units in Umea, Sweden, in August 2006. This activity was quite challenging as several important problems should be solved. As the bioagent simulants are disseminated in air during the trial, the detector should be linked to the cyclone device collecting particles into buffer and transferring the captured sample to the detector. An overall view of the testing arrangement is available in Figure 6.

Furthermore, the detector must operate in an isolated manner during the trial field some hundred meters away from the personnel safely protected from the clouds of simulants. The communication was realized using TCP/IP transfer of the serial data through a local network, the wireless communication was not allowed due to military regulations. In fact, several biosensor-based instruments were controlled through common internet data links [xiv], using LabView-based program modules [xv]. The optical biosensor Analyte 2000 was successfully used on a small Swallow unmanned aircraft with complete ground-based control and evaluation of results [xvi].

Participation of our ImmunoSMART device in the field trial brought to us completely new experience not achievable in laboratory conditions. The ImmunoSMART system appeared rugged enough to survive the transportation without any problems and it was immediately ready for operation. The next step, connection of our device with the cyclone air sampler provided by organizers of the trial was considered as a simple task, however, on site it was realized that the cyclone was not operating in a constant manner, i.e. the flow of sample was often irregular and contained a lot of air bubbles disturbing the detector part. The optimal setting of three pumps in the cyclone was time consuming, affected by temperature changes due to weather fluctuations, and cyclone settings could not be controlled remotely.

Figure 6. Field testing of ImmunoSMART (right, placed inside a briefcase) linked to a cyclone module (BioTrace, left, provided by FOI Umea, Sweden) sampling the surrounding air space for bacterial simulants. The system is operated from a distant location using internet link as a carrier for the serial communication with computer (RS232-TCP/IP module EC232, www.papouch.com).

The sampling was carried out only once during the dissemination event (20 min, totally spread either 4 g of dried agents or 500 ml of 10^7 to 10^9 CFU/ml of suspensions), the bioagent simulants (*E. herbicola, B. subtilis* and *globigii*, ovalbumin) in the flow from cyclone was directed to the measuring cell, washed, incubated with tracer and the response was measured amperometrically with substrates. Each sensor was used only once and discarded after the measurement. The main problems realized during the field testing included:

- low specificity of assays (significant cross-reactivities due to partially optimized assay conditions and compromised selection of antibodies, resulting from the short preparation time)
- negative effect of the detergent present in the working solution of the cyclone

- device malfunction – circuitry for one of the pumps failed and it was not possible to exchange this part on site (the substrate mixture was used instead of mixing both components before the measuring cell)
- instability of the substrate mixture (hydrogen peroxide and iodide) under temperature fluctuations resulting in high background currents
- lack of a reference method applicable for confirmation of bioagents in the sample from the cyclone (though agar plates were commonly used to quantify bacterial particles in the air during dissemination)

To summarize, this first participation in the field trial was not very successful. On the other hand, the obtained experience was invaluable and such practical problems can not be simulated in the strictly controlled laboratory conditions and become discovered only in real life situations. The good point was that the detector functioned well during the complete two-week testing period, with only the failure of the pump as mentioned above. The mechanical construction of the device should be modified in order to substantially improve resistance to humidity condensation and rain (solved by placing in a closed briefcase container) and also to allow decontamination of the exposed surface (protection of connectors). The external measuring parts (immunosensor and the cell) should be more compact, closed and tightly fixed to the detector body, and also the containers with working solutions should be better secured.

A major change might be the addition of a switching valve to improve flexibility of the flow-through system; this was originally not planned due to quite large dimensions of common electrically actuated valves and high demands on power. Finally, the regeneration of the immunosensor should be re-considered in order to allow longer repeated operation without manual manipulations with immunosensors.

Conclusions

The development of the amperometric immunosensor device planned for bioagents detection was briefly described with results from the initial testing of the device in real situations. The transfer of the detector system from the laboratory to the real world demonstrated several more or less significant problems which together made the function of the detector rather unreliable. However, the experience gained during this phase of technology transfer from basic research approaches to practical evaluations under unpredictable field conditions is invaluable and it will help to correct several weak parts of the detector. Finally, one last problem should be solved; so far, the conclusion whether the target bioagent was detected or not, is manually determined by the user looking at trace signal measurements. This evaluation and decision-making should be implemented in the control software, and this might be quite a challenging task.

Acknowledgements

Financial resources for this research were provided by the Ministry of Defense of Czech Republic. Dr. Inga Gustafson and Dr. Torbjorn Tjarnhage from FOI, Umea, Sweden, are gratefully acknowledged for generous help during the field trial and for providing the cyclone device.

References

[i] D.V. Lim, J.M. Simpson, E.A. Kearns, M.F. Kramer, Current and developing technologies for monitoring agents of bioterrorism and biowarfare, *Clin. Microbiol. Revs.* **18** (2005), 583.

[ii] J.J. Gooding, Biosensor technology for detecting biological warfare agents: Recent progress and future trends, *Anal. Chim. Acta* **559** (2006), 137-151.

[iii] B.M. Paddle, Biosensors for chemical and biological agents of defense interest, *Biosens. Bioelectron.* **11** (1996), 1079-1113.

[iv] S.S. Iqbal, M.W. Mayo, J.G. Bruno, B.V. Bronk, C.A. Batt, J.P. Chambers, A review of molecular recognition technologies for detection of biological threat agents, *Biosens. Bioelectron.* **15** (2000), 549-578.

[v] F.S. Ligler, Biosensors for identification of biological warfare agents, *Biosens. Bioelectron.* **14** (2000): 749-749.

[vi] A.A. Bergwerff, F. Van Knapen, Surface plasmon resonance biosensors for detection of pathogenic microorganisms: Strategies to secure food and environmental safety, *J. AOAC Int.* **89** (2006), 826-831.

[vii] F.S. Ligler, K.E. Sapsford, J.P. Golden, L.C. Shriver-Lake, C.R. Taitt, M.A. Dyer, S. Barone, C.J. Myatt, The array biosensor: Portable, automated systems, *Anal. Sci.* **23** (2007), 5-10.

[viii] C.C. Jung, E.W. Saaski, D.A. McCrae, B.M. Lingerfelt, G.P. Anderson, RAPTOR: A fluoroimmunoassay-based fiber optic sensor for detection of biological threats, *IEEE Sens. J.* **3** (2003), 352-360.

[ix] V.R. Rivera, F.J. Gamez, W.K. Keener, J.A. White, M.A. Poli, Rapid detection of *Clostridium botulinum* toxins A, B, E, and F in clinical samples, selected food matrices, and buffer using paramagnetic bead-based electrochemiluminescence detection, *Anal. Biochem.* **353** (2006), 248-256.

[x] O.A. Sadik, W.H. Land, J. Wang, Targeting chemical and biological warfare agents at the molecular level, *Electroanal.* **15** (2003), 1149-1159.

[xi] J. Shah, E. Wilkins, Electrochemical biosensors for detection of biological warfare agents, *Electroanal.* **15** (2003), 157-167.

[xii] J.P. Hart, A. Crew, E. Crouch, K.C. Honeychurch, R.M. Pemberton, Some recent designs and developments of screen-printed carbon electrochemical sensors/biosensors for biomedical, environmental, and industrial analyses, *Anal. Let.* **37** (2004), 789-830.

[xiii] P. Skládal, Y. Symerská, M. Pohanka, B. Šafář, A. Macela, Electrochemical immunosensor for detection of *Francisella tularensis*. In: Defense against Bioterror: Detection Technologies, Implementation, Strategies and Commercial Opportunities, Eds.: D. Morrison, F. Milanovich, D. Ivnitski, T.R. Austin, *NATO Security through Science Ser. B – Physics and Biophysics*, Springer, Dordrecht, The Netherlands, 2005, pp. 221-232.

[xiv] J. Tschmelak, G. Proll, J. Riedt, J. Kaiser, P. Kraemmer, L. Barzaga, J.S. Wilkinson, P. Hua, J.P. Hole, R. Nudd, M. Jackson, R. Abuknesha, D. Barcelo, S. Rodriguez-Mozaz, M.J.L. de Alda, F. Sacher, J. Stien, J. Slobodnik, P. Oswald, H. Kozmenko, E. Korenkova, L. Tothova, Z. Krascsenits, G. Gauglitz, Automated water analyser computer supported system (AWACSS) Part II: Intelligent, remote controlled, cost-effective, on-line, water-monitoring measurement system, *Biosens. Bioelectron.* **20** (2005), 1509-1519.

[xv] K.S. Chang, H.D. Jang, C.F. Lee, Y.G. Lee, C.J. Yuan, S.H. Lee, Series quartz crystal sensor for remote bacteria population monitoring in raw milk via the Internet *Biosens. Bioelectron.* **21** (2006), 1581-1590.

[xvi] G.P. Anderson, K.D. King, D.S. Cuttino, J.P. Whelan, F.S. Ligler, J.F. MacKrell, C.S. Bovais, D.K. Indyke, R.J. Foch, Biological agent detection with the use of an airborne biosensor, *Field Anal. Chem. Technol.* **3** (1999), 307-314.

L.M. Lechuga et al. (Eds.)
IOS Press, 2008

Disposable Screen Printed Electrochemical Sensors and evaluation of their application as Alarm Systems against Terrorism

Serena LASCHI, Ilaria PALCHETTI, Marco MASCINI
Dipartimento di Chimica, Università di Firenze
e-mail:marco.mascini@unifi.it

Introduction

The most frequently used methods for the unambiguous identification of Chemical Warfare Agents (CWAs), their precursor and breakdown products are based on gas chromatography (GC) in combination with mass spectrometry (GC-MS) and/or tandem mass spectrometry (GC-MS/MS), liquid chromatography (LC) coupled with MS (MS), and nuclear magnetic resonance (NMR) spectrometry.
However, there are increasing needs for a portable, end-user friendly instrument for rapid (seconds or minutes), sensitive determination of CWAs. Sensor and Biosensor technologies are claimed to satisfy these requirements, and especially electrochemical based sensors.

Electrochemical sensors represent the most rapidly growing class of (bio)-chemical sensors[1]. Compared to other sensors, electrochemical sensors are especially attractive because of their remarkable detection capability, experimental simplicity and low cost. They have a leading position among the presently available sensors that have reached the commercial stage and which have found a vast range of important applications in the fields of clinical, industrial and biomedical analyses [2]. Recent developments in some electrochemical methods (such as square wave voltammetry, stripping voltammetry and chronopotentiometric stripping analysis) have opened the door for the application of electrochemical sensors to the analysis of more chemical species.

Between electrochemical sensors, planar devices have a large development in latest years and they have been brought into use for measuring different applications [1,3,4]. Production techniques used for the development of planar electrochemical sensors can be classified in different ways; as other 'Solid-state sensors', they have been made from classical semiconductor materials (such as Si or Ge), solid electrolytes, insulators, metals and catalytic materials; thick film and thin film technologies, photolithography and silicon technology are the most used techniques to produce planar electrochemical sensors.

Electrochemical sensors respond also to the most important trend in the research of latest years, which is to go smaller in device dimensions [5]; micro- and nanoelectrodes and microtechnological approaches to manufacturing sensors demonstrate the strong link between chemistry, physics, and engineering in this field. Finally, the integration into microfluidic platforms is just the latest, logical step in the miniaturization direction, especially for clinical applications [5].

The current paper will focus attention on the recent developments in thick film electrochemical sensors, showing also some of their possible applications as alarm systems against terrorism.

1. Screen-Printed Electrodes production

Thick-film technology was introduced about thirty years ago as a means of producing hybrid circuits. A key factor distinguishing a thick-film circuit is the method of film deposition, namely screen printing, which is one of the oldest forms of graphic art reproduction. The use of thick-film techniques shows several significant advantages respect to other techniques, such as flexibility of design and choice of materials, low cost in infrastructure, possibility of automation of the fabrication process and low barrier for technological transfer capability [8,9]. A thick-film sensor is generally formed by layers (or films) of special inks or pastes deposited sequentially onto an insulating support or substrate [10,11]. The film is applied through a mask contacting the substrate, and deposited films are obtained by pattern transfer from the mask (Fig. 1).

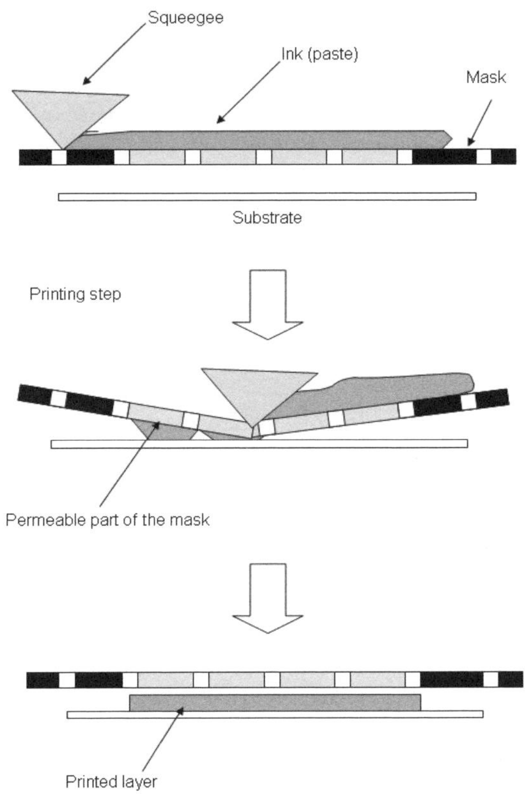

Fig. 1: A scheme of the different steps of the screen printing process

After the printing step, the next stage is to dry the printed film. All pastes contain various organic solvents which are needed in order to produce the correct viscosity for screen printing; these can be removed by drying the film in a conventional box oven. Conventionally, thick-film sensors were baked at temperatures ranging from 300 and 1200 ◦C; alternatives include cold-cured ink formulations and a photocured process using ultraviolet light. After drying, the films retain a rigid pattern on the substrate and are relatively immune to smudging. Complex structures can be built up by repeating the print process using the materials appropriate to the specific design and a range of mask designs. In this way, a variety of screen-printed thick-film devices can thus be fabricated (Fig. 2). One of major advantages of this technology is the ability to produce hybrid integrated circuits in a robust and miniaturized package. These planar devices present in fact many advantages including disposability and small dimensions, which facilitates the design of portable measuring systems.

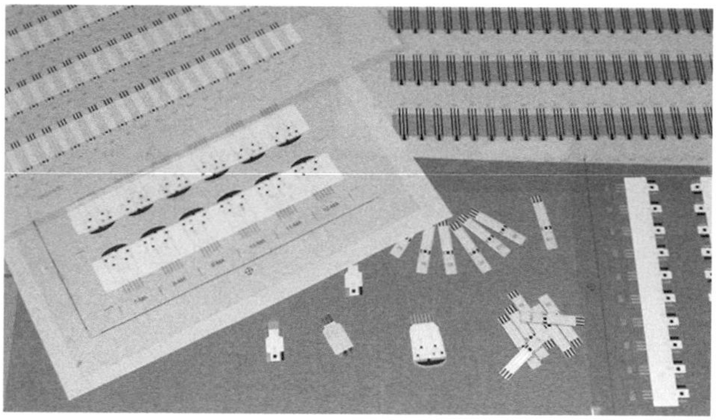

Fig. 2: A picture showing various screen-printed electrodes

2. Screen-Printed Biosensor

Screen printing technique is also one of the most promising technologies to produce planar electrochemical biosensors [12–14] to be placed large-scale on the market in the near future, because of advantages such as miniaturization, versatility and low cost and particularly for the possibility of mass production [15].

The most well-known application of screen-printing technology in biomedical application has been in the clinical analysis of blood glucose levels in people with diabetes [16]. The deceptively simple combination of a fungal enzyme (glucose oxidase, GOD) with an electrochemical detector as a screen-printed electrode has effectively met the needs of the 1–2% of the world's population that have diabetes. Enzyme-based electrode biosensors have been used to test glucose levels in blood samples since 1975, but the emergence of a convenient, hand-held commercial format revolutionized their use [17]. Many examples of enzyme based biosensor using screen printed electrodes are nowadays reported in literature.

Another interesting approach in the use of screen-printed electrodes is based on the use of bioaffinity molecules to obtain disposable affinity biosensors. In the case of antibodies immobilization, the electrode constitutes both the solid-phase for the immunoassay and the electrochemical transducer, giving rise to an immunosensor. A recent example of a disposable immunosensor (based on enzymatic labeling of one of the reagents) to detect polychlorinated biphenyls (PCBs) was reported in [20]. In this case, the surface of the electrode was modified using different strategies to produce two different formats of immunochemical tests based on indirect and competitive assays. The graphite surface of the screen-printed electrodes was modified by immobilization of antibodies or antigens in order to obtain a selective surface for the realization of the immunological reaction.

In the family of affinity biosensors, screen-printed DNA biosensors are also included. DNA biosensors present an enormous potential for early clinical diagnosis of genetically inherited diseases, on site detection of food contamination, forensic studies and environmental monitoring. DNA biosensors involve the use of nucleic acids as biological recognition elements to explore novel hybridization probes, transduction strategies and potential practical applications. Electrochemical techniques have been used to differentiate between ss (prehybridized) and ds (hybridized) DNA using several approaches. Farabullini et al. [21] report the construction of disposable electrochemical biosensor array for simultaneous detection of food borne pathogens. The latter DNA biosensor was realized by immobilizing calf thymus DNA at a fixed potential onto an electrode surface. The calf thymus DNA sensor was then immersed in a sample solution containing the analytes. After 2 min of interaction a chronopotentiometric analysis (PSA) was carried out to evaluate the oxidation of guanine residues on the electrode surface. In this case, it was possible to evaluate the electrochemical effects resulting from the presence of genotoxic compounds as 2-aminoanthracene, estradiol or bisphenol A [22].

3. Screen-Printed Biosensor as Alarm system against terrorism

Chemical Warfare Agents (CWAs) can be divided into several groups, of which the most important are nerve agents, blistering agents or vesicants, blood agents and incapacitating agents [23, 24]. Nerve agents derive their name from their adverse effects on the nervous system. These are all organophosphates that contain: a phosphonate group that mimics the acetate moiety of the acetylcholine (ACh) molecule; an ester or thioester linkage; and often a positively charged group mimicking the choline moiety. Many commonly used agricultural pesticides are organophosphate compounds (e.g. chlorpyrifos, parathion) [23, 25-26].

The principal effect of nerve agents is the inhibition of the enzyme acetylcholinesterase (AChE), which is essential for terminating the action of the neurotransmitter ACh. The phosphonate moiety of nerve agents mimics the acetate group of acetylcholine; unlike ACh, however, the nerve agent is capable of modifying the AChE protein in either a slowly reversible or irreversible fashion. ACh is a neurotransmitter at the neuromuscular junction, in the parasympathetic nervous system and in a number of brain regions. Actually, nerve agent intoxication results in an accumulation of endogenous acetylcholine and continual stimulation of the nervous system. One nerve agent in particular, sarin, was in the news after its use against the population of the Kurdish village of Birjinni in 1993 and after a terrorist attack in the Tokyo underground system on 20 March 1995. Another well-known nerve agent is VX, which is among the most toxic substances ever produced by man [22].

In this review, recent advancements in the author lab for the detection of nerve agents in the form of AChE inhibitors will be described.

The results were obtained using a model compound, i.e. carbofuran. The measurement of the anticholinesterase activity of this compound, a well known AChE inhibitor, can be used to perfom a screening test to evaluate the presence of AChE inhibitors in different matrices. The inhibition was evaluated using screen-printed electrodes in chrono-amperometric measurements.

It should be noted that in the author's lab different strategies have been developed for this purpose [27-30]; however, in the current paper, we want to describe an interesting method based on the use of a modern electroanalytical technique, i.e. chronoamperometry that allows rapid measurement with a very good sensitivity.

Thus, screen-printed carbon electrodes (SPCEs) were modified by incorporating in the ink an optimized percentage of cobalt(II) phthalocyanine (CoPC), an electrochemical mediator. As reported in literature, among the electrochemical mediators, CoPC was indicated as one of the most suitable for the detection of thiol-containing molecules and the resulting oxidation signals occur at lower voltages, thus limiting the electrochemical interference of other oxidizable compounds. Using these modified SPCEs, under optimized chronoamperometric conditions, it was possible to detect AChE inhibitors, such as Carbofuran, through the study of the AChE activity. Actually, the AChE free in solution was incubated with the pesticide. The inhibitory effect of the pesticide determines a decrease of the catalytic activity of AChE. As a consequence, less thiocholine is produced and thus a current value lower than that recorded in a blank solution was obtained. This current decrease was correlated with the pesticide concentration. A detection limit of 2.0×10^{-10} M for Carbofuran was found in an analysis time of 15 min. (fig.3).

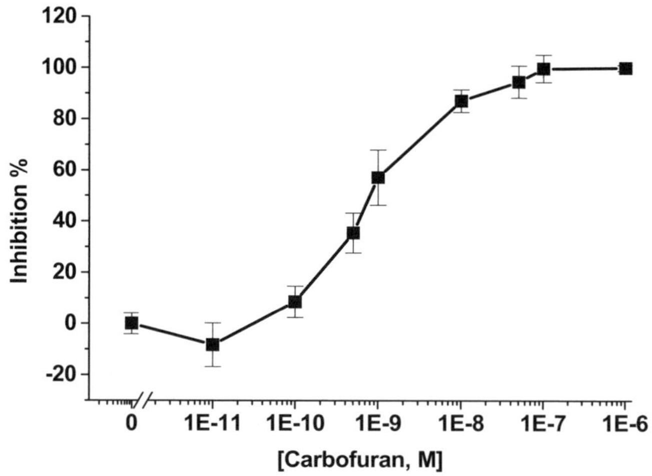

Fig. 3: Inhibition plot of Carbofuran obtained by using CoPC-modified SPCEs coupled with AChE free in solution. Measurements were performed by chronoamperometry. Applied potential +0.1 V. Other conditions are reported in the text.

In order to increase the versatility of the device, a reproducible and reliable immobilization strategy of AChE onto the SPCE surface was studied. The AChE was immobilized by cross-linking with glutaraldehyde, Bovine Serum Albumin (BSA) and Nafion® onto the surface of the modified SPCE. The composition of the surface protein layer (enzyme units, glutaraldehyde, BSA and Nafion® amounts) was optimized to obtain high and reliable response towards the substrate and AChE inhibitors. In the optimized conditions, the dynamic range for Carbofuran detection was 10-10 to 10-7 M with a detection limit of 4.9 × 10-10 M, for an analysis time of 15 min. (fig. 4). This is an important feature, considering that the immobilization can determine a loss of the activity of the enzyme that influence the sensitivity as well as dynamic range of the pesticide detection. Moreover, the proposed method was less prone to electrochemical interferences since the incubation and measurement were performed in two separate steps.

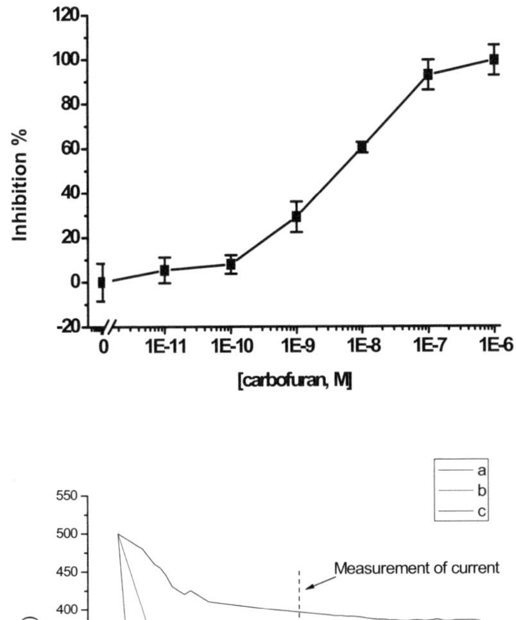

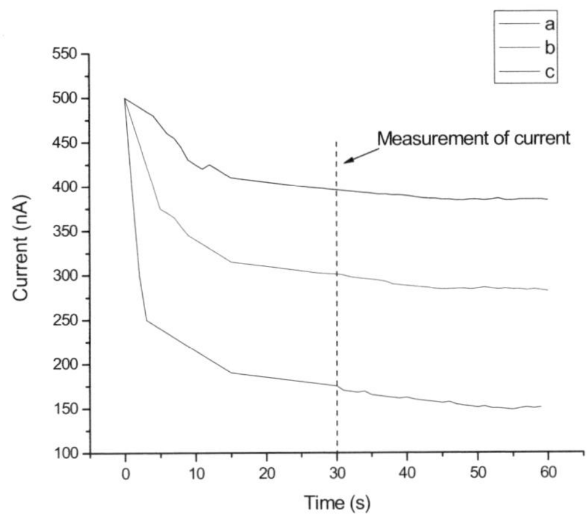

Fig. 4: a) Inhibition plot of carbofuran onto AChE-based biosensor, b) Typical chronoamperograms obtained after incubation with different concentration of Carbofuran: (a) 0 M; (b) 10^{-9} M (c) 10^{-7}M.

The modification of a SPCE with CoPC provided significant improvement in the performance of the AChE biosensor. This was revealed in the decrease of the working potential of the TCh oxidacoordination and in a fast and reliable response towards the specific substrate, ATCh. The limits of detection obtained are small enough to detect trace amounts of anticholinesterase compounds. Taking into account the simplicity of the proposed method, the device offers good opportunities for direct and inexpensive detection of toxic contaminants in field and as alarm system against terrorism.

Conclusion and Future Perspective

Screen-printed electrodes coupled to biomolecules and modern electroanalytical techniques offer an interesting opportunity for the development of sensitive, rapid low cost analytical method for warfare agents. Method based on AChE inhibitors was already demonstrated.

However, in the near future another interesting class of receptors for biological warfare agent detections will be more and more used, i.e. aptamers.

Aptamers specific for these particular targets, such as anthrax spores, cholera toxin, staphylococcal enterotoxin B, ricin and abrin toxin, have been selected in the last years and, by using aptamers, different detection systems have been developed.

Aptamers are single stranded DNA or RNA ligands which can be selected for different targets starting from a huge library of molecules containing randomly created sequences. Aptamers have been selected to bind very different targets, from proteins to small organic dyes.

In addition to the very important aspect of having an unlimited source of identical affinity recognition molecules available due to the selection process, aptamers can offer advantages over antibodies that make them very promising for analytical applications.

The possibility to couple aptamer with electrochemical transduction was recently demonstrated, as well as the possibility to use screen-printed as electrochemical transducers [31].

References

[1] Stetter JN, Penrose WR, Sheng Y. Sensors, chemical sensors, electrochemical sensors, and ECS. J Electrochem Soc 2003;150:S11–6.
[2] Edmonds TE. Chemical sensors. New York: Chapman and Hall; 1998.
[3] Mortimer RJ, Beech A. AC impedance characteristics of solid-state planar electrochemical carbon monoxide sensors with Nafion as solid polymer electrolyte. Electrochim Acta 2002;47:3383–7.
[4] Lauks IR. Microfabricated biosensors and microanalytical systems for blood analysis. Acc Chem Res 1998;31:317–24.
[5] Bakker E, Telting-Diaz M. Electrochemical sensors. Anal Chem 2002;74:2781–800.
[6] Prudenziati M, editor. Thick-film sensors. Amsterdam: Elsevier; 1994.
[7] White NM, Turner JD. Thick-film sensors: past, present and future. Meas Sci Technol 1997;8:1–20.
[8] Galan-Vidal CA, Munoz J, Dominguez C, Alegret S. Glucose biosensor strip in a three electrode configuration based on composite and biocomposite materials applied by planar thick film technology. Sens Actuators B 1998;52:257–63.
[9] Alvarez-Icaza M, Bilitewski U. Mass production of biosensors. Anal Chem 1993;65:525A–33A.
[10] Wang J, Pamidi PVA. Disposable screen printed electrodes for monitoring hydrazines. Talanta 1995;42:463–7.
[11] White NM, Turner JD. Thick-film sensors: past, present and future. Meas Sci Technol 1997;8:1–20.

[12] Albareda-Sirvent M, Merkoci A, Alegret S. Configurations used in the design of screen-printed enzymatic biosensors. A review. Sens Actuators B 2000;69:153–63.
[13] Palchetti I, Cagnini A, Del Carlo M, Coppi C, Mascini M, Turner APF. Determination of anticholinesterase pesticides in real samples using a disposable biosensor. Anal Chim Acta 1997;337:315–21.
[14] Mulchandani A, Chen W, Mulchandani P,Wang J, Rogers KR. Biosensors for direct determination of organophosphate pesticides. Bios Bioelectron 2001;16:225–30.
[15] Lucarelli F, Kicela A, Palchetti I, Marrazza G, Mascini M. Electrochemical DNA biosensor for analysis of wastewater samples. Bioelectrochem 2002;58:113–8.
[16] Wang J, Chen Q. Enzyme microelectrode array strips for glucose and lactate. Anal Chem 1994;66:1007–11.
[17] Cass AEG, Davis G, Francis GD, Hill HAO, Aston WJ, Higgins IJ, et al. Anal Chem 1984;56:667.
[18] Higgins I.J., Hill H.A.O. and Plotkin E.V. 1985. US Patent 4545382.
[19] Lam Y-Z, Atkinson J. Disposable screen-printed biosensor for transcutaneous oxygen measurement. Meas Sci Technol 2002;13: 2074–81.
[20] Laschi S, Franek M, Mascini M. Screen-printed electrochemical immunosensors for PCB detection. Electroanalysis 2000;12:1293–8.
[21] Farabullini F., Lucarelli F., Palchetti I., Marrazza G., Mascini M., Disposable Electrochemical Genosensor for the Simultaneous Analysis of Different Bacterial Food Contaminants, Biosensors & Bioelectronics, (2007) 22, 1544-1549.
[22] Lucarelli F, Palchetti I, Marrazza G, Mascini M. Electrochemical DNA biosensor as a screening tool for the detection of toxicants in water and wastewater samples. Talanta 2002;56:949–57.
[23] Hooijschuur E W.J., Hulst A.G. de Jong A. L., de Reuver L. P., van Krimpen S.H., van Baar B.L.M., Wils E.R.J., Kientz C.E., Brinkman U.A.Th., Identification of chemicals related to the chemical weapons convention during an interlaboratory proficiency test, TRAC, 21, 2, 2002, 116-130.
[24] Bismuth C., Borron S.W., Baud F J., Barriot P., Chemical weapons: documented use and compound on the horizon, Toxicology Letters 149 (2004) 11-18
[25] Noort D., Benschop H.P., Black R.M., Biomonitoring of exposure to chemical warfare agents: A review, Toxicology and Applied Pharmacology, 184, 116-126 (2002)
[26] Sanchez-Santed F., Canada F., Flores P., Lopez-Grancha M., Cardona D., Long-term neurotoxicity of paraoxon and clorpyrifos: behavioral and pharmacological evidence, Neurotoxicology and teratology, 26, 2004, 305-317.
[27] A. Cagnini, I. Palchetti, I. Lionti, M. Mascini, A.P.F. Turner, Disposable ruthenized screen-printed biosensors for pesticides monitoring, Sensors and actuators B 24-25, (1995), 85-89.
[28] A. Cagnini, I. Palchetti, M. Mascini, A.P.F. Turner, Ruthenized Screen-printed Choline Oxidase-Based Biosensors for measurement of anticholinesterase Activity, Mikrochim. Acta 121, 155-166 (1995).
[29] I. Palchetti, A. Cagnini, M. Del Carlo, C. Coppi, M. Mascini, A.P.F. Turner, Determination of anticholinesterase pesticides in real samples using a disposable biosensor, Analytica Chimica Acta, 337 (1997) 315-321.
[30] S. Hernandez, I. Palchetti, M. Mascini, Determination on Anticholinesterase activity for pesticides monitoring using Acetylthiocholine Sensor, Int. Journal of Environmental Analytical Chemistry, 78, 3-4, (2000), 263-278
[31] S. Centi, S. Tombelli, M. Minunni, M. Mascini, Aptamer-Based Detection of Plasma Proteins by an Electrochemical Assay Coupled to Magnetic Beads, Anal. Chem. 2007; 79(4); 1466.

L.M. Lechuga et al. (Eds.)
IOS Press, 2008

New Generation Biosensors Based on Direct Bioelectrocatalysis and Multi-Microchannel Technology

Dmitri IVNITSKI[1], Plamen ATANASSOV[1], Brittany BRANCH[1],
Christopher APBLETT[2] and Vladimir SHAPOVALOV[3]
[1]University of New Mexico, ABQ, USA
[2]Sandia National Laboratories, ABQ, NM, USA
[3]*MER Corporation, Tucson, AZ, USA*

Abstract: Biological warfare agents are the most problematic weapons of mass destruction and terror. Both civilian and military sources predict that over the next decade the threat from proliferation of these agents will increase significantly. Therefore, the ability to accurately predict the dispersion, concentration, and ultimate fate of biological warfare agents released into the environment in real time is essential to prepare for and respond to a biological warfare agent release. A fusion of micro- and nanotechnologies with biosciences could significantly counter biological threat agents on the battlefield. Miniaturization of biosensor technologies has great potential for improving resolution time (speed of assay), reducing reagent use, and allowing for higher sample throughput. Fast analysis and on-chip integration of supporting electronic circuitry for signal analysis and remote control would enable sensing at a remote location. This paper describes a new biosensor technology based on combination of direct bioelectrocatalysis and multi-microchannel technology. To demonstrate direct electron transfer, glucose oxidase and PQQ-dependent glucose dehydrogenase have been selected. An electrochemical sensor, which includes biological sensing element immobilized on the surface of microchannels of a working electrode, can be used in the form of a flow-through amperometric, potentiometric, or conductometric device.

Keywords. Biosensor, Direct Electron Transfer, Multi-Microchannel Technology

Introduction

The effective testing of biological threat agents requires portable sensor technology, which should be extremely sensitive, universal, reliable, and fast. A biosensor should be miniaturized, use fewer consumables, and be of low maintenance relative to current equipment that has been deployed for "real time" monitoring of biological warfare agents. The sensors must be able to detect biological agents at threshold concentrations in a minimum of 5-10 minutes and capable of being used anytime, and anywhere.

In current biosensor technology (second generation) artificial redox mediators are used to shuttle electrons between the active site of the enzyme and the electrode surface [1-3]. Application of redox mediators allows a significant decrease in the applied potential and it can minimize interference with coexisting electroactive compounds present in real samples. But, the stability and toxicity of some mediators limit their application. The leakage of immobilized mediators out of the electrode surface rapidly decreases the biosensor response. Therefore, in recent years biosensors not requiring mediators have been receiving intense attention [4-6]. Biosensors of 3rd generation, in which enzymes can operate in the absence of any mediators have an excellent potential for miniaturization and to create extremely selective, sensitive and simple bioanalytical systems. The near perfect selectivity for the target analyte can be achieved through a design, in which the biosensor can operate in the absence of any mediators. Direct method of assay offers the opportunity to obtain biological information, such as the kinetics and mechanism of biospecific binding, epitope mapping, steric and allosteric effects. The mediatorless biosensors based on direct electron transfer (DET) between the active site of the enzyme and the electrode surface can work in a potential range close to the redox potential of the enzyme itself therefore, becoming less exposed to interfering reactions. This approach has an excellent potential for miniaturization and to create extremely selective, sensitive and simple bioanalytical systems.

There are several critical challenges which must be overcome for the successful application of DET approach for biosensor technology. A major challenge is the fact that the electron-transfer rate between the active site of large redox proteins and the electrode is usually slow because the redox center of most oxidoreductases is located deeply in the apoenzyme. For example, the depth of the redox center for GOx is approximately 13 – 15 Å. There are different promoters which can be used to reduce the electron tunneling distance between the active site of the enzyme and the electrode surface. For example, the application of nanoparticles, which can be incorporated between the initial donor and the final acceptor sites (i.e., a donor-bridge-acceptor system). Such configurations are referred to as a biological "wire" chain (see Fig. 1.a). The experimental evidence of DET has been demonstrated for both low molecular weight electron-transfer proteins and enzymes with large complicated structures [7-14].

Mediatorless Biosensor Based On Direct Electron Transfer (DET)

To demonstrate DET, we have selected two enzymes, glucose oxidase (GOx) and PQQ-dependent glucose dehydrogenase (GDH). Both enzymes have highly catalytic activity high stability and are available and rather inexpensive, and are widely used for design and development of glucose sensors. The key tasks in the design and development of biosensors based on DET are to build highly porous three-dimensional nanoelectrode networks on the micro-electrode surface and to immobilize enzyme molecules inside of such three-dimensional nano-electrode structures. In this work, multi-walled carbon nanotubes (MWNTs) were grown directly on the Toray® carbon paper (TP) [14]. The incorporation of the MWNTs in TP was provided by the chemical vapor deposition technique after electrochemical deposition of cobalt nanoparticles as a catalyst for carbon nanotubes (CNT). The combination Toray® carbon electrode with carbon nanotubes provides for a highly porous three-dimensional network with increased electrode surface area from the original 2 m^2/g to two up to orders of magnitude. This and provides efficient electrical communication between the active site of enzyme and the electrode surface.

The experimental data shows that both enzymes, GOx and GDH, immobilized on the CNTs surface demonstrate direct electron transfer combined with high biocatalytic activity to glucose. Typical cyclic voltammograms of GOx electrode in the absence and in the presence of glucose are shown in Figure 1.b.

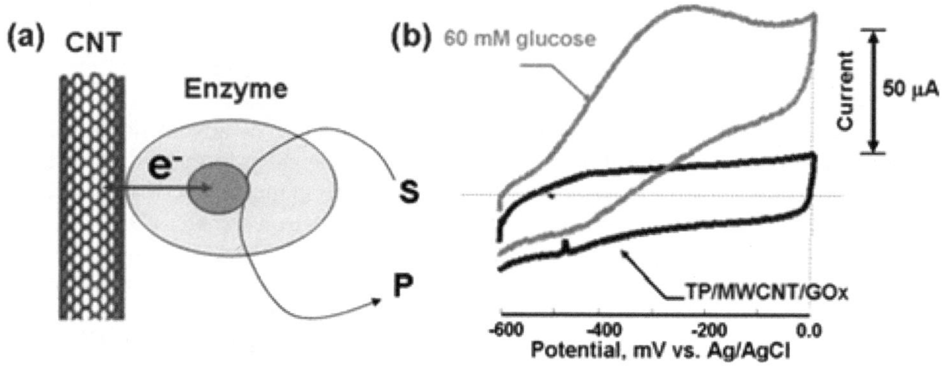

Figure 1. "Wired-enzyme electrode" concept: direct electron transfer between the active site and the carbon nanotube (a). Cyclic voltammograms of GOx/CNT/C electrode in the absence (1) and in the presence of 20 mM glucose. Scan rate is 50 mV/s (b)

A large electrocatalytic anodic current is observed in the presence of glucose and in the absence of any redox mediators. The CNTs play an active role concerning direct electron transfer between the active site of the enzyme and the electrode surface. Enzymatic electrodes can operate with redox potentials close to the active site of the enzyme. This statement was supported by potentiometric and amperometric data (Figs 2a and 2b).

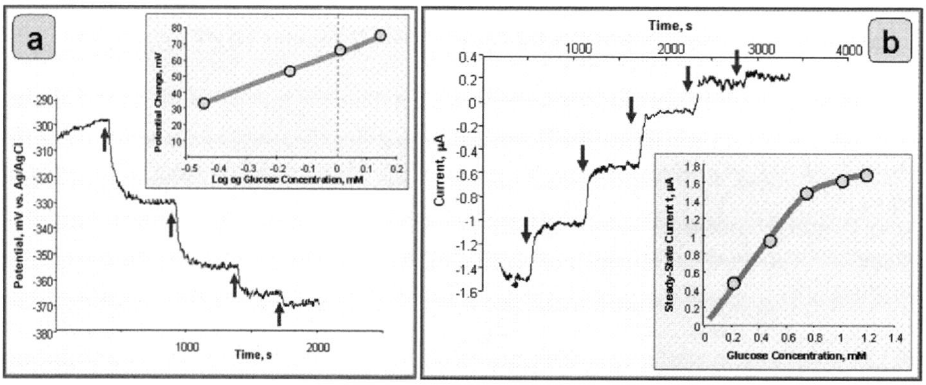

Figure 2. Potentiometric responses of GOx/CNTs/C electrode at different glucose concentrations. Inset: plot of the potentiometric responses as a function of glucose concentration (a). Amperometric responses upon incremental successive glucose addition. Inset is the plot of the steady-state current vs. glucose concentration (b).

In the presence of glucose redox potentials are shifted immediately and reach a stable negative magnitude between -380 mV and -400 mV vs. Ag/AgCl at concentrations of glucose 1.5 mM and higher. There are no potential changes observed with GOx/C electrodes without carbon nanotubes. The amperometric measurements were provided at an applied potential of -0.2 V. We found that analytical signals immediately changed in the presence of glucose and, there are no current changes in the absence of CNTs on the electrode surface. The glucose biosensor based on direct electron transfer can reach up to 95% of the steady-state current in less than 2 seconds (response time) where the sensitivity of the biosensor is 2 μA/mM. This shows that enzyme biosensors based on direct electrical communication approach have the follow advantages:

- Direct signal transduction
- Ideal approach for miniaturization
- Reagentless, fast, selective, robust and flexible platform
- Simple sensor calibration
- Wide range of applications

Three-Dimensional Multi Microchannel Technology

Recent advances in microarray technology, microelectromechanical systems (MEMS), microfluidics, microseparations and optoelectronics present new technological possibilities for producing fast, extremely sensitive and inexpensive "smart" sensing systems for field applications [15-25]. Miniaturization of biosensor technologies has intrinsic advantages for improving resolution time (speed of assay), reducing reagent use, and allowing for higher sample throughput. A fusion of micro- and nanotechnology with biology has great potential for the development of low-cost disposable chips for rapid molecular analysis that can be carried out with simple handheld devices. It has been shown that conducting chemical or biological reactions in ultra narrow microchannels or porous materials allows a significant reduction of transport limitations across the unstirred layer (Nernst diffusion layer). Microanalytical devices have found many applications, ranging from the life sciences industries for pharmaceuticals and biomedicine (drug design, delivery and detection, diagnostic devices) to industrial applications of combinatorial synthesis (such as rapid chemical analysis and high throughput screening). Miniaturization of biosensors into a single integrated “lab-on-a-chip” system possesses great potential for environmental monitoring, point-of-care testing, and food analysis, which includes high sensitivity, improved accuracy, lower power and sample consumption, disposability and automation. An emerging demand is to monitor and detect chemical and biological warfare agents in real-time.

Most approaches developed to date are based on the application of two dimensional microchip formats, wherein a suitable set of biological receptor elements (enzyme, antibody, DNA, protein, etc.) are immobilized on the surface of a planar microchip substrate. The diffusion-controlled rate of biospecific reactions are accelerated in microfluidic systems through the use of microchannels or porous substrates that provide a unique means to prepare a three-dimensional network suited for the immobilization of different biomolecules. However, in two dimensional microchip formats the density of receptor spots is ultimately limited by either the dispensing mechanism or the amount of biological recognition material within each spot. This fact negatively impacts the dynamic range and lower detection limit of analysis. The sensitivity of electrochemical detection based on microelectrodes is typically substantially lower then conventional techniques. Currently, a single microchannel with special inner geometry is individually wired and used in microchip technology. But so far there is a large uncertainty in the test results because of variations in the properties of individual microchannels, i.e., there is no reproducibility of test results. Improvements have been reported through the use of porous substrates [21-25]. The porous substrates provide a three-dimensional hydrophilic environment similar to free solution for biomolecular interactions.

Recently, we have applied a new multi-microchannel technology [26] for the design and development the three-dimensional working electrodes with immobilized biological sensing elements for the detection and identification of different biological active agents in real-time. The method of preparation of a three-dimensional multi-microchannel substrate includes the following steps (Figs. 3a and 3b).

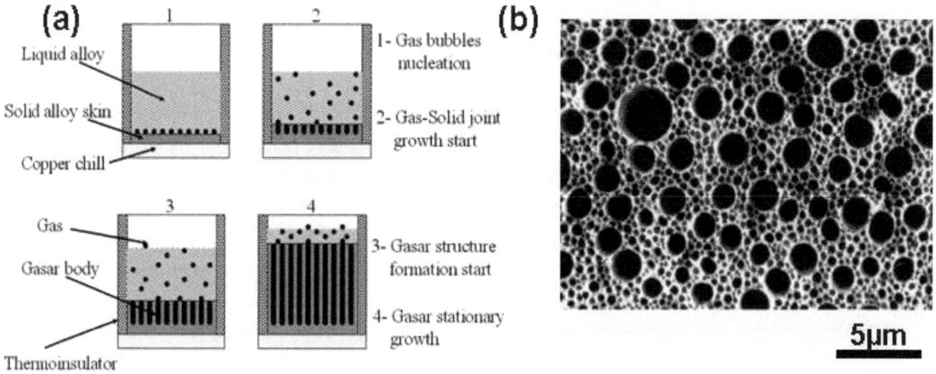

Figure 3. Schematic sketch preparation of multi-microchannel substrate (a) and a sample that have been prepared from gold metal (b).

The metal is melted in an autoclave in an atmosphere of hydrogen under a partial pressure between 2 and 5 atmospheres in order to dissolve the gas in the melt. After saturating, the melt is cooled resulting in the solubility of the dissolved gas is sharply decrease. The gas bubbles grow in parallel with solidification while producing microchannels. The main benefits of this multi microchannel technology are:
Sensitivity of assay is achieved due to large surface area to volume ratio, larger binding capacity and shorter hybridization times

Speed of assay is significantly accelerated due to enhanced mass transport within the channels. Analytical signal is directly proportional to the electrode surface, concentration and the diffusion coefficient and inversely proportional to the thickness of the Nernst diffusion layer.

Ideal material for miniaturization and development reagentless bioanalytical approaches

A wide variety of metals and other materials can be used for development of multi-microchannel flow-through working electrodes. For example Fe, Ni, Co, Cr, Cu, Mg, Mo, W, Al, Au, Ir, Ru, Pd, Pt, Zr, Ti, Rh, alloys thereof, ceramic and glass. The multi-microchannel element produced in this approach can be used as an electrochemical sensor, micromixer, or a fuel element. An electrochemical sensor which includes biological sensing element immobilized on the surface of microchannels of a working electrode can be used in the form of a flow-through amperometric, potentiometric, or conductometric device.

The surface of microchannels serves as both the solid phase for immobilization of the biorecognition agents as well as a transducer. Figure 4 presents principal components of the Flow Biochip incorporating multi-microchannel electrodes.

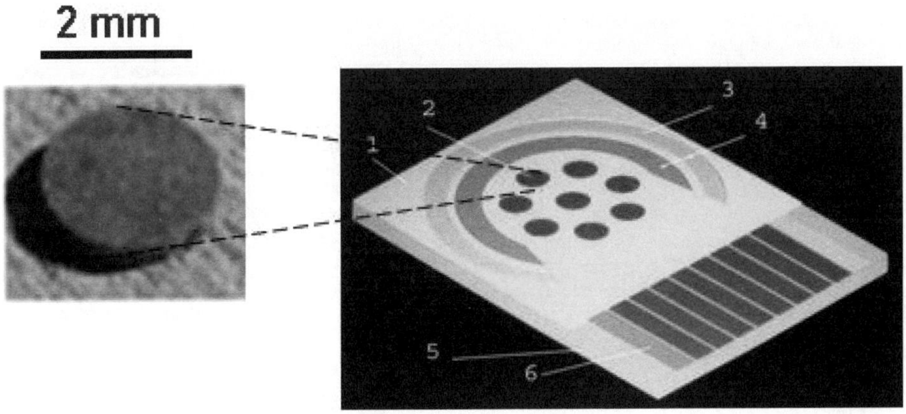

Figure 4. Schematic of a Flow Biochip composed of eight circular stainless steel multi-microchannel working electrodes (2, and the insert photograph), flat stainless steel counter (3) and Ag/AgCl (4) reference electrodes.

Each working electrode contains a high density of non-interconnected parallel microchannels with a diameter between 2 μm and 3 μm. The Biochip was tested for detection and identification of *Campylobacter* species [26]. Lectins have been used as biorecognition element for bacteria. They interact with bacteria by binding to the carbohydrate structures presented at the cell surface. The operation of the Flow Biochip is based on transmembrane ion-current modulation caused by specific interaction of bacteria with immobilized lectins [26].

The combination of direct bioelectrocatalysis approach with a multi-microchannel technology opens wide perspectives for the development of a new generation of micro scale biosensors with near perfect selectivity for target analytes.

References

[1] Dennis Morrison, Fred Milanovich, Dmitri Ivnitski, and Thomas Austin (eds). (2005). Defense Against Bioterror: Detection Technologies, Implementation Strategies and Commercial Opportunities. NATO Science Series, Kluwer Academic Publishers, Boston/Dordrecht/London, 400 p.

[2] Walt, D., and D.R. Franz. 2000. Biological warfare detection. Anal Chem. 72: 739A-746A.

[3] Iqbal, S.S., M.W. Mayo, J.G. Bruno, B.V. Bronk, C.A. Batt, and J.P. Chambers. 2000. A review of molecular recognition technologies for detection of biological threat agents. Biosens. Bioelectron. 15: 549-578.

[4] Zhang, W., Fan, C., Sun, Y., Li, G. 2003. An electrochemical investigation of ligand binding abilities of film-entrapped myoglobin. Biochimica et Biophysica Acta, 1623, 29– 32

[5] Dai Z., Yan F., Chen J., Ju H. 2003. Reagentless Amperometric Immunosensors Based on Direct Electrochemistry of Horseradish Peroxidase for Determination of Carcinoma Antigen-125, Anal. Chem., 75, 5429-5434

[6] J.-F. Rochettea, E. Sachera, M. Meuniera, J.H.T. Luong. 2005. A mediatorless biosensor for putrescine using multiwalled carbon nanotubes, Analytical Biochemistry, 336, 305–311
[7] J. Zhang, M. Grubb, A.G. Hansen, A.M. Kuznetsov, A. Boisen, H. Wackerbarh, J. Ulstrup, Electron transfer behavior of biological macromolecules towards the single-molecule level. J. Phys.: Condens. Matter, 15 (2003) S1873-S1890.
[8] E. Katz, A.N. Shipway, I. Willner, In: G. Schmid (Ed.), Nanoparticles - From Theory to Applications, Wiley-VCH, Weinheim, Germany, 2004, Ch. 6, 368.
[9] G.S. Sayler, M.L. Simpson, C.D. Cox, Emering foundations: nano-engineering and bio-microelectronics for environmental biotechnology. Curr. Opin. Microbiol., 7 (2004) 267.
[10] K. Kano, T. Ikeda, Bioelectrocatalysis, powerful means of connecting electrochemistry to biochemistry and biotechnology. Electrochemistry, 2003, 71, N2, 86-99
[11] S. Shleev, J. Tkac, A. Christenson, T. Ruzgas, A. I. Yaropolov, J. W. Whittaker, Lo Gorton, Direct electron transfer between copper-containing proteins and electrodes. Biosensors and Bioelectronics, 20 (2005) 2517–2554.
[12] G. Presnova, V. Grigorenko, A. Egorov, T. Ruzgas, A. Lindgren, Lo Gorton, T. Börchersc, Direct heterogeneous electron transfer of recombinant horseradish peroxidases on gold. Faraday Discuss., 2000, 116, 281-289.
[13] T. Ikeda, F. Fushimi, K. Miki, M. Senda, Direct bioelectrocatalysis at electrode modified with D-gluconate dehyrogenase, Agric. Biol. Chem., 52 (1988) 2655-2658.
[14] D. Ivnitski, B. Branch, P. Atanasov, C. Apblett, Glucose oxidase anode for biofuel cell based on direct electron transfer. 2006. Electrochemistry Communications, 8, 1204-1210.
[15] Min, J.H., and A. Baeumner. 2003. The micro-total analytical system for the detection of bacteria/viruses. J. Ind. Eng. Chem. 9: 1-8.
[16] Stenger, D.A., J.D. Andreadis, G.J. Vora, and J.J. Pancrazio. 2002. Potential applications of DNA microarrays in biodefense-related diagnostics. Curr. Opin. Biotech. 13: 208-212.
[17] Rowe, C.A., L.M. Tender, M.J. Feldstein, J.P. Golden, S.B. Scruggs, B.D. MacCraith, J.J. Cras, and F.S. Ligler. 1999. Array biosensor for simultaneous Identification of bacterial, viral, and protein analytes. Anal. Chem. 71: 3846-3852.
[18] McDonald, R., T. Cao, and R. Borschel. 2001. Multiplexing for the detection of multiple biowarfare agents shows promise in the field. Mil. Med. 166: 237-239.
[19] D. Ivnitski, D. O'Neil, A. Gattuso, R. Schlicht, M. Moore, M. Calidonna, R. Fisher. (2003). Nucleic acid approaches for detection and identification of biological warfare and infectious disease agents, BioTechniques, 35, 862-869.
[20] Stratis-Cullum, D.N., G.D. Griffin, J. Mobley, A.A. Vass, and T. Vo-Dinh. 2003. A miniature biochip system for detection of aerosolized *Bacillus globigii* spores. Anal. Chem. 75: 275-280.
[21] Benoit, V., A. Steel, M. Torres, Y.-Y. Yu, H. Yang, and J. Cooper. 2001. Evaluation of three-dimensional microchannel glass biochips for multiplexed nucleic acid fluorescence hybridization assays. Anal. Chem. 73: 2412-2420.
[22] Schöning, M.J., A. Kurowski, M. Thust, P. Kordos, J.W. Schultze, and H. Lüth. 2000. Capacitive microsensors for biochemical sensing based on porous silicon technology. Sensor. Actuat. B-Chem. 64: 59-64.
[23] Kricka, L.J. 2001. Microarrays, biochips and nanochips: personal laboratories for the 21st Century. Clin. Chem. Acta 307: 219–223.
[24] Kopf-Sill, A.R. 2002. Success and challenges of lab-on-a-chip, Lab Chip, 2: 42N-47N.
[25] Steel, A., M. Torres, J. Hartwell, Y-Y. Yu, N. Ting, G. Hoke, and H. Yang. 2000. The flow-thru chip: a three-dimensional biochip platform, p. 87-117. *In* M. Schena (Ed.), Microarray Biochip Technology, A BioTechniques Books Publication, Eaton Publishing, CA.
[26] D. Ivnitski, V. Shapovalov, Method of producing a multi-microchannel flow-through element and device using same. US Patent Application, 20070034298, 2007.

Commercial and Pre-Commercial Cell Detection Technologies for Defence against Bioterror
L.M. Lechuga et al. (Eds.)
IOS Press, 2008

Electro-optical analysis as a tool for determination of microbial cells with the help of specific bacteriophages

Oleg V. IGNATOV[a1], Olga I. GULIY [a], Viktor D. BUNIN [b]
[a]*Institute of Biochemistry & Physiology of Plants & Microorganisms, Russian Academy of Sciences, Russian Federation*;
[b] *SRC Applied Microbiology and Biotechnology, Obolensk, Russian Federation*

Abstract. This article describes electro-optical (EO) characterization of biospecific binding between *Escherichia coli* XL-1 and phage M13K07. The electro-optical analyzer (ELUS EO), which has been developed at the State Research Center for Applied Microbiology, Obolensk, Russia, was used as the basic instrument for electro-optical measurements. The operating principle of the analyzer is based on the polarizability of microorganisms, which depends strongly on their composition, morphology, and phenotype. The principle of analysis of the interaction of *E. coli* with phage M13K07 is based on recording changes in the optical parameters of bacterial suspensions. The phage–cell interaction includes the following stages: phage adsorption on the cell surface, entry of viral DNA into the bacterial cell, amplification of phage within the infected host, and phage ejection from the cell. In this work, we used M13K07, a filamentous phage of the family *Inoviridae*. Preliminary study had shown that combination of the EO approach with a phage as a recognition element has excellent potential for mediatorless detection of phage–bacteria complexes. The interaction of *E. coli* with phage M13K07 induced a strong and specific electro-optical signal as a result of substantial changes in the EO properties of the *E. coli* XL-1 suspension infected by phage M13K07. The signal was specific in the presence of foreign microfloras (*E. coli* K-12 and *Azospirillum brasilense* Sp7). Integration of the electro-optical approach with a phage has the following advantages: (1) bacteria from biological samples need not be purified, (2) the phage infection of bacteria is specific, (3) exogenous substrates and mediators are not required for detection, and (4) it is suitable for any phage–bacterium system when bacteria-specific phages are available.

Keywords: *Escherichia coli* XL-1, electro-optical spectrum; phage M13K07, transfection, bacteria–phage binding

[1]Corresponding author: Dr. Oleg V. Ignatov, Institute of Biochemistry and Physiology of Plants and Microorganisms, Russian Academy of Sciences, 13 Prospekt Entuziastov, Saratov 410049, Russian Federation. Tel.: +7(8452)970444, Fax: +7(8452)970383, E-mail: oignatov@ibppm.sgu.ru

Introduction

Detection and identification of pathogenic bacteria in the environment present multiple challenges [1, 2]. First, pathogenic agents are effective in extremely small numbers and, therefore, biodetection systems need to exhibit high sensitivity. Second, the complex and rapidly changing environmental background and the fact that many pathogenic organisms differ little from the normal flora require detection systems to exhibit a high degree of selectivity. Detection systems must be able to discriminate biological agents from other harmless biological and nonbiological material present in the environment. Third, another challenge that needs to be addressed is speed. Unlike chemical agents, many living biological agents can reproduce, multiply inside the host, and be passed from person to person. Therefore, rapid detection and identification of biological agents are crucial. Most of the current biodetectors employ either nucleic acid or immunological recognition methods combined with optical, electrochemical, or mass transduction [1–5]. An advantage of immunological methods with regard to the nucleic acid-based analysis is that a cell/spore lysis step is eliminated [6]. However, the use of antibodies in the latter methods as sensing elements of biosensors has a number of drawbacks. Antibodies are not sufficiently stable, the quality of antibodies can vary with different animals and production variables and they are expensive. For example, polyclonal antibodies obtained from immunized animals are not selective, since they recognize all antigens to which the animal has been exposed in the past. Monoclonal antibodies are more selective, but their application is limited by their inherited sensitivity to unfavorable environmental conditions. Therefore, considerable effort is now directed towards the development of new analytical approaches for pathogen detection.

We recently reported on a new electro-optical analyzer (ELUS) for electrical manipulation and identification of cells [7–11]. The operating principle of the analyzer is based on the polarizability of microorganisms, which depends strongly on their composition, morphology, and phenotype. The benefits of the analyzer have been demonstrated for studying electrophysical parameters, structures, and population heterogeneity of microorganisms in the absence and presence of toxic low-molecular weight agents [7–11]. The theoretical and experimental aspects of the electro-optical approach have been developed in a number of laboratories [12–14].

The current focus of our research group is to apply the electro-optical analyzer to direct detection of and discrimination between bacteria on the basis of biospecific binding. More specifically, the purpose of this paper is to demonstrate the potential of using the electro-optical technique in combination with bacteriophage amplification for direct monitoring of *E. coli*. Bacteriophages are virus particles that generally attach to and infect a narrow range of host cells [15–18]. Infection of bacteria by a bacteriophage starts by recognition of the host through binding to an outer membrane receptor. Like human viruses, bacteriophages inject their genetic material into the bacterial cell, replicate by hundreds per cell, and then burst out before moving on to the next host cell. In the case of tailed phage, this binding triggers conformational changes that are transmitted along the tail to the capsid, allowing its opening and the release of the viral genome, which causes a change in the dielectric properties of the cells. Advantages of the phage technology include its simplicity, ease of use, low cost, safety, and immunogenicity (for vaccine studies). The bacteriophage specificity has been demonstrated at both the species and the strain levels in the literature. Specific bacteriophages are very good indicators for determining the species and type of bacteria. That is why they have found a wide application in medical practice for the fast identification of viable bacteria [16, 18, 19]. The phage–cell interaction is fairly complex and depends on the structure of the phage itself [20–21] and the presence of F-pili in bacteria [22]. In our work, *Escherichia coli* XL-1 and bacteriophage M13K07 were chosen as a model system. Bacteriophage M13 is an *E. coli*-specific filamentous phage. It is a long, thin bacterial virus that infects *E. coli* cells without cell lysis [23, 24]. The class I filamentous bacteriophage M13 includes circular single-stranded DNA enclosed in a cylindrical proteinic envelope (length, ca. 890 nm; diameter, ca. 7 nm). The phage M13 infection of *E. coli* cells containing F-pili has been extensively studied [25, 26].

Experimental

Microorganisms

Escherichia coli XL-1 was obtained from the culture collection held at the Institute of Biochemistry and Physiology of Plants and Microorganisms, RAS, Saratov.

Culture conditions

E. coli XL-1 was grown in a liquid medium containing (g l^{-1}): NaCl, 10; yeast extract (FLUKA, Switzerland), 5; peptone (FLUKA, Switzerland), 5. The cultures were shaken (160 rpm) aerobically for 24 h at a constant temperature of 30°C. The grown cells were used for EO studies.

Cell preparation

Cells to be used in the analysis were washed three times by centrifugation at 2800 × *g* for 5 min and were resuspended in distilled water to an optical density OD_{665} of 0.4–0.5. To remove cellular aggregates, we recentrifuged the cell suspension at 110 × *g* for 1 min, and further work was carried out with the suspension that remained in the supernatant liquid. Cell suspensions were prepared so that the OD_{665} for each type of cell suspension was within the range of 0.4–0.42.

Measurement of the electrorotation spectra (ES)

The electro-optical analyzer (ELUS EO), developed at the State Research Center for Applied Microbiology, Obolensk, Russia, and was used as the basic instrument for electro-optical measurements. The analyzer consists of the following modules: a unit for sample preparation, a mixer, an AC field generator, an EO flow cell, a microcontroller of liquid stream transactions, a thermal system, an operator interface, and an image processor. The sample preparation unit includes a hydraulic system, containers for the storage of initial ingredients, a reaction vessel, a filter module, and a peristaltic pump for sampling. Vacuum is used as a driving force. The analyzer operates at multiple software programmable frequencies ranging from 0.4 kHz to 20 MHz. A software program was used for collection, calculation, and processing of data. The orientation spectra (OS) of the cells were measured with the EO ELUS analyzer at a wavelength of 670 nm. The measurements were performed with a discrete set of frequencies of the orienting electric field (250, 500, 1000, and 2000 kHz).

Monitoring of the interaction of E. coli with phage M13K07

The principle of analysis of the interaction of *E. coli* with phage M13K07 is based on recording changes in the optical parameters of bacteria by using an image processing technique. The assay includes measurement of the EO spectrum of *E. coli* in the absence and presence of the phage. The phage–cell interaction includes the following stages: phage adsorption on the cell surface, entry of viral DNA into the bacterial cell, amplification of phage within the infected host, and phage ejection from the cell. In this work, we used M13K07, a filamentous phage of the family *Inoviridae*. M13 K07, a kanamycin-resistant commercial preparation manufactured by Stratagene (Sweden), was constructed on the basis of wild-type phage M13 [27]. The phage has specificity toward *E. coli* XL-1. For transfection, *E. coli* XL-1 and TG-2 (a separate colony) were transferred from a plate containing agar-supplemented LB medium with 12.5 μg/ml tetracycline to a plate containing 2 ml of LB medium. The culture was incubated overnight with constant aeration at 35^{0}C; then, 1/10th of the overnight culture was transferred to a fresh medium of the same composition and was grown to exponential phase with aeration at 37^{0}C. When cells reached early log phase (OD_{660} = 0.5—0.6, corresponding to 7.7×10^{8} cells/ml), the aeration was stopped for 30–40 min so that the cells could restore their F-pili, and the suspension was incubated in a thermostat at 37^{0}C. The concentrations of the organisms present were checked by standard techniques with the help of light microscopy. Twenty phage per bacterium were used for infection. Upon direct addition of the phage, the culture was incubated at 37^{0}C in a thermostat without shaking, in order for phage particles to sorb at the surface of the pili. After that, the cells were prepared for EO measurements as described earlier. As a result of formation of a microorganism–phage complex, the optical properties of the cell suspension, particularly light scattering properties, are changed and lead to changes in optical density.

Results and discussion

As is known, the infection of *E. coli* male cells by bacteriophage M13, fd, or f1 begins with the interaction of the minor phage capsid protein g3p (gene 3 protein) with bacterial F-pili (which is the primary phage receptor) and, subsequently, with the integral membrane protein TolA [24]. Phage action on a bacterial cell may follow different paths: a lytic reaction, lysis from the outside, and a lysogenic reaction. Whatever the path, due to entry of viral DNA into the *E. coli* cell cytoplasm, with the capsid protein (g8p) integrated into the inner cytoplasmic membrane, the phage infection leads to considerable changes in the electrophysical parameters of the cells, including changes in their dielectric properties and various kinds of damage to the intracellular structures [7–14, 25]. Therefore, discrimination of microorganisms might be done by measuring the electro-optical spectra of the microorganisms in the absence and presence of phage. For controlling phage transfection to the bacteria, we grew the cells in an LB nutrient medium containing kanamycin, because phage M13K07 has resistance to this antibiotic [27]. The cells grew well with kanamycin, indicative of phage transfection. **Figure 1** shows the electro-optical spectra of *E. coli* after incubation with various numbers of phage M13K07.

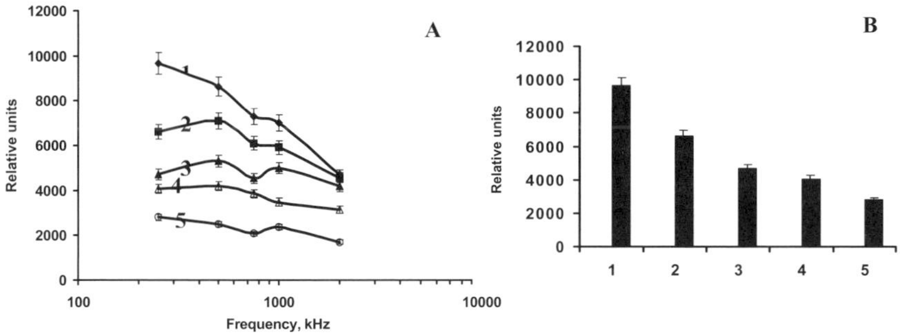

Figure. 1. (A) The electro orientation spectra and (B) relative units at 250 kHz of viable *E. coli* XL-1 cells suspended in distilled water (conductivity, 1.8 μS/m), obtained after incubation with phage M13K07. (1) control without phage; (2) 1 phage per bacterium; (3) 5 phage per bacterium; (4) 10 phage per bacterium; (5) 20 phage per bacterium.

As a result of these studies, we showed that the maximum OS changes occurred when the cells were infected at a rate of twenty (20) phage per bacterium. To record cell infection, we added 20 phage per bacterium to the suspension in the subsequent experiments. In fact, we found that even a ratio of one phage per bacterium led to changes in the EO spectrum. But the best resolution was obtained at a ratio of 20 phage per bacterium.

The next parameter studied was the effect of the time of infection of the *E. coli* XL-1 cell suspension by phage M13K07 on the electro-optical signal. The data are presented in **Figure 2**.

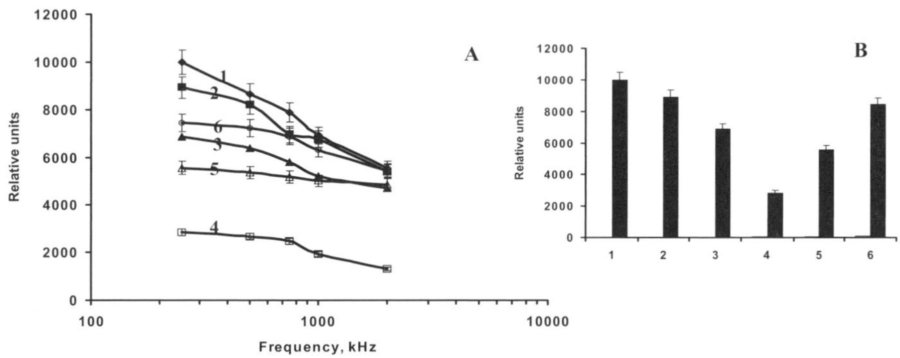

Figure. 2. (A) The electro orientation spectra and (B) relative units at 250 kHz of viable *E. coli* XL-1 cells suspended in distilled water (conductivity, 1.8 μS/m), obtained after incubation with 20 M13K07 phage per bacterium: (1) control without phage; (2) after 1 min; (3) after 10 min; (4) after 30 min; (5) after 60 min; (6) after 90 min.

Considerable changes in the magnitude of the EO signal were detected in the first 5 min after the phage injection. The reason might possibly be the formation of a bacteria–phage complex. These results are in harmony with the data in the literature, because the translocation of phage DNA occurs with the participation of the bacterial proteins TolQ, TolR, and TolA after the pili supposedly are retracted, thereby transferring the phage to the bacterial surface [26]. After cell–phage incubation for 60 min, the magnitude of the cell-suspension EO signal increased considerably. This may have been due to the entry of DNA into the cell and to the processes occurring in the cytoplasm. One major process is the penetration of single-stranded phage DNA into the bacterial-cell cytoplasm and its transformation into a double-stranded, replicate form, which is the basis for the further synthesis of all phage proteins [26].

Endemann and Model [24] showed that before assembly of the bacteriophage, all its minor capsid proteins are integral proteins of the *E. coli* inner membrane. Consequently, all structural and morphogenetic proteins of the bacteriophage are localized in the infected-cell membrane, compatible with the model according to which phage assembly occurs concurrently with phage ejection from the cell. Thus, the substantial increase in the magnitude of the EO signal after cell–phage incubation for 90 min may be explained by the possible assembly of the phage and their egress from the cell.

Since many pathogenic organisms differ little from the normal flora, it is important to test the specificity of the electro-optical system in the presence of interfering factors, first and foremost in the presence of a foreign microflora. Therefore, in our next experiments the electro-optical measurements were done in the absence and presence of a foreign microflora that would not be infected by phage M13K07 (**Figure 3**).

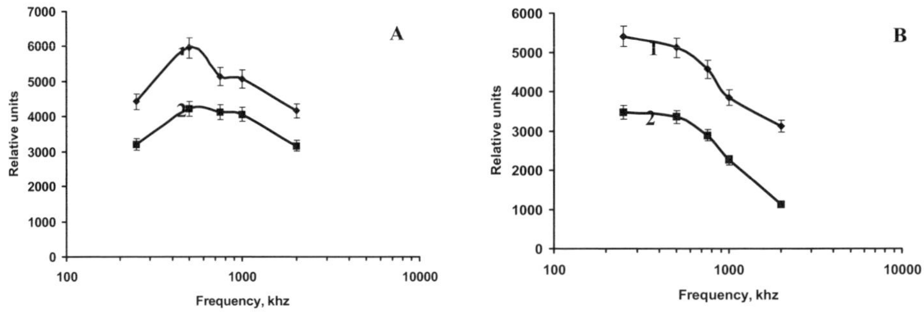

Figure. 3. The electro orientation spectra of mixed viable cells (*E. coli* XL-1 and *E. coli* K-12) (A) and (*E. coli* XL-1 and *A. brasilense* Sp7) (B) suspended in distilled water (conductivity, 1.8 μS/m), obtained after incubation with 20 M13K07 phage. (1) control without phage; (2) 20 phage per bacterium.

As a control, we used cells of *E. coli* K-12 and *Azospirillum brasilense* Sp7. *E. coli* K-12 was chosen because it is a parent strain of *E. coli* XL-1 and can possibly be infected by M13K07. *A. brasilense* Sp7 was chosen because it occupies a different taxonomic position and has a cell size similar to that of *E. coli* XL-1. To this end, the phage was added to a mixed suspension containing *E. coli* XL-1 and K-12 (**Fig**. 3A), and *E. coli* XL-1 and *A. brasilense* Sp7 (**Fig**. 3B) (OD_{665}, 0.42–0.44). The cell–phage interaction conditions were the same as those used in the experiments with XL-1 alone. We found that during the formation of an *E. coli* XL-1–phage M13K07 complex in the presence of the foreign microflora (both *E. coli* K-12 and *A. brasilense* Sp7), there occurred a substantial decrease in the magnitude of the EO signal (**Figure** 4). Control experiments were run in parallel to explore the possibility of nonspecific interaction of phage M13K07 with *E. coli* K-12 and *A. brasilense* Sp7. To test this assumption, we infected an *E. coli* K-12 suspension with phage M13K07 at a rate of 20 phage per bacterium. The electro-optical signal did not change on addition of the phage; that is, the phage did not infect K-12 (**Fig**. 4A). When *A. brasilense* Sp7 was incubated with the phage, the cell OS did not change either; that is, the phage did not infect the cells (**Fig.** 4B). Thus, we found that specific changes in the EO parameters of the cell suspensions under the influence of M13K07 occur only in *E. coli* XL-1 and do not occur in K-12 or in *A. brasilense* Sp7.

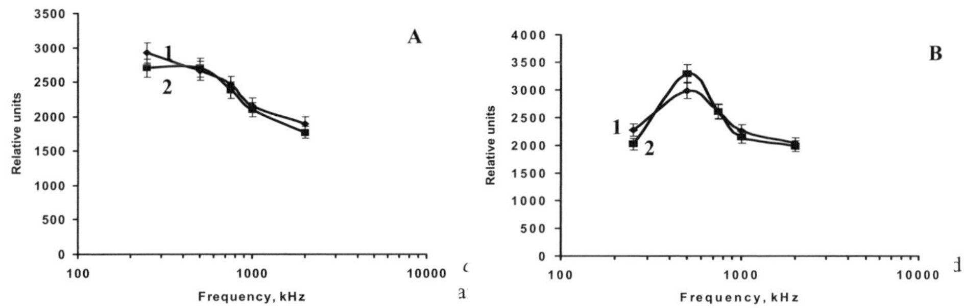

Figure 4. The electro orientation spectra of viable *E. coli* K-12 (A) and *A. brasilense* Sp7 (B) cells suspended in distilled water (conductivity, 1.8 μS/m), obtained after incubation with 20 M13K07 phage. (1) control without phage; (2) 20 phage per bacterium.

Conclusions

This preliminary study has shown that combination of the EO approach with a phage as a recognition element has excellent potential for mediatorless detection of phage–bacteria complexes. The interaction of *E. coli* with phage M13K07 induces a strong EO signal as a result of substantial changes in the EO properties of the *E. coli* XL-1 suspension infected by phage M13K07. This approach has the following advantages: (1) bacteria from biological samples need not be purified, (2) the phage infection of bacteria is specific, (3) exogenous substrates and mediators are not required for detection, and (4) it is suitable for any phage–bacterium system when bacteria-specific phages are available.

Combination of the EO approach with the phage technology is a generic technology that enables rapid and specific detection of viable bacteria and might be a basis for development of portable biosensor-based systems for detection of pathogens in medical research, food processing, and environmental analysis.

Acknowledgements

This work has benefited from grants of the *Fond sodeistviya otechestvennoi nauke* (the National Science Support Foundation); the President of the Russian Federation (no. MK-8599.2006.4); State Support for the Leading Scientific Schools of the Russian Federation (no. NSh-6177.2006.4); ISTC (no. 3170PDG); the *Fundamental'nye nauki – meditsine* (Basic Sciences for Medicine) Basic Research Program of the Presidium of the Russian Academy of Sciences; and the Russian Foundation for Basic Research (no. 07-04-00301).

References

[1] S.S. Iqbal, M.W. Mayo, J.G. Bruno, B.V. Bronk, C.A. Batt, and J.P. Chambers, A review of molecular recognition technologies for detection of biological threat agents, *Biosensors & Bioelectronics* **15** (2000), 549-578.

[2] D. Ivnitski, I. Abdel-Hamid, P. Atanasov, E. Wilkins, (Review), Biosensors for detection of pathogenic bacteria, *Biosensors & Bioelectronics* **14** (1999), 599-624.

[3] P.B. Luppa, L.J. Sokoll, D.W. Chan, Immunosensors-principles and applications to clinical chemistry (Review), *Clinica Chimica Acta* **314** (2001), 1 – 26.

[4] D.M. Norton, Polymerase chain reaction-based methods for detection of Listeria monocytogenes: Toward real-time screening for food and environmental samples, *J. AOAC Int.* **85** (2002), 505-515.

[5] J.H. Min and A. Baeumner, The micro-total analytical system for the detection of bacteria/viruses, *J. Ind. Eng. Chem.* **9** (2003), 1-8.

[6] D.N. Stratis-Cullum, G.D. Griffin, J. Mobley, A.A. Vass, and T. Vo-Dinh, A miniature biochip system for detection of aerosolized *Bacillus globigii* spores, *Anal.Chem.* **75** (2003), 275-280.

[7] V.D. Bunin, A.G. Voloshin, Determination of cell structures, electrophysical parameters, and cell population heterogeneity, *J. Colloid Interface Sci.* **180** (1996), 122-126.

[8] V.D. Bunin, A.G. Voloshin, Z.F. Bunin, V.A. Shmelev, Electrophysical monitoring of culture process of recombinant *Escherichia coli* strains, *Biotechnol. Bioengineer* **51** (1996), 720-724.

[9] O.V. Ignatov, S .Yu. Shchyogolev, V.D. Bunin, V.V. Ignatov, Electro-physical properties microbial cells during aerobic metabolism toxic compounds. In: Biotransformations: Bioremediation Technology for Health and Environment Protection. Edited by: Ved Pal Singh & Raymond D. Stapleton, Elsevier Science B.V., The Netherlands, **36** (2001), 403-425

[10] O.V. Ignatov, O.I. Guliy, S.Yu. Shchyogolev, V.D. Bunin, V.V. Ignatov, Effect of *p*-nitrophenol metabolites on microbial-cell electro-optical characteristics, *FEMS Microbiol. Lett.* **214** (2002), 81-86.

[11] O.I. Guliy, O.V. Ignatov, S.Yu. Shchyogolev, V.D. Bunin, V.V. Ignatov, Quantitative determination of organophosphorus aromatic nitro insecticides by using electric-field cell orientation in microbial suspensions, *Analytica Chimica Acta* **462** (2002), 165-177.

[12] A. Miroshnikov, V.M. Fomchenkov, and A.Yu. Ivanov, *Electrophysical Analysis and Cell Separation,* Nauka Publishers, Moscow, 1986, p. 40-65 [in Russian].

[13] J. Gimsa, D. Wachner, A unified resistor-capacitor model for impedance, dielectrophoresis, electrorotation, and induced transmembrane potential, *J. Biophys.* **75** (1998), 1107–1116.

[14] P. Gascoyne, R. Pethig, J. Satayavivad, FF. Becker, M. Ruchirawat,. Dielectrophoretic detection of changes in erythrocyte membranes following malarial infection, *Biochim. Biophys. Acta* **1323** (1997), 240-252.

[15] G. Smith, V. Petrenko, Phage display, *Chem Rev.* **97** (1997), 391-410.

[16] V.A. Petrenko, V.J. Vodyanoy, Phage display for detection of biological threat agents, J. *Microbiol. Met.* **53** (2003), 253-262.

[17] I. Benhar, Biotechnological applications of phage and cell display, *Biotechnology Advances* **19** (2001), 1-33.

[18] E.R. Goldman, M.P. Pazirandeh, J.M. Mauro, K.D. King, J.C. Frey, G.P. Anderson, Phage-displayed peptides as biosensor reagents, *J. Molecular Recognition* **13** (2000), 382-387.

[19] S. Chatterjee, M. Mitra, S. Gupta, A high yielding mutant of mycobacteriophage L1 and its application as a diagnostic tool, *FEMS Microbiology Letters* **188** (2000), 47-53.

[20] I. Stengele, P.Bross, X. Garces, J. Giray, I. Rasched, Dissection of functional domains in phage fd adsorption protein. Discrimination between attachment and penetration sites, *J. Mol. Biol.* **212** (1990), 143.

[21] K.S. Jakes, N.G. Davis, N.D. Zinder, A hybrid toxin from bacteriophage f1 attachment protein and colicin E3 has altered cell receptor specificity, *J. Bacteriol.* **170** (1988), 4231.

[22] M. Russel, H. Whirlow, T.P. Sun, R.E. Webster, Low-frequency infection of F-bacteria by transducing particles of filamentous bacteriophages, *J. Bacteriol.* **170** (1988), 5312

[23] S.A. Overman, M. Tsuboi, G.J. Thomas, Subunit orientation in the filamentous virus Ff (fd, f1, M13), *Journal of Molecular Biology* **259** (1996), 331 – 336.

[24] H. Endemann, P. Model Location of filamentous phage minor coat proteins in phage and in infected cells, *J. Mol Biol* **250** (1995), 496 – 506.

[25] L.W. Deng, P. Malik, R.N. Perham, Interaction of the globular domains of pIII protein of filamentous bacteriophage fd with the F-pilus of *Escherichia coli*, *Virology* **253** (1999) 271.

[26] E.M. Click, R.E. Webster, Filamentous phage infection: required interactions with the TolA protein, *J. Bacteriol.* **179** (1997), 6464-6471.

[27] H. R. Hoogenboom, A.D. Griffits, K.S. Johnson, D.J. Chiswell, P. Hundson, G. Winter, Multi-subunit proteins on the surface of filamentous phage: methodologies for displaying antibody (FAB) heavy and light chains; *Nucleic Acids Res.* **19** (1991), 4133-4137.

L.M. Lechuga et al. (Eds.)
IOS Press, 2008

Fast Measurement of Cells Status by Electro-optical Technique

Victor BUNIN[1,2], Alexander ANGERSBACH[2]
[1] SRC Applied Microbiology and Biotechnology, Obolensk, Russia
[2] Biotronix GmbH, Henningsdorf, Deutschland

Abstract. Determination of bacterial cells status during its identification is the important part of their selective analysis. Heterogeneity of cells ensemble with identical superficial properties, but a different metabolic activity, is typical for a bacteria in the nature. Electro optical phenomenon was used for cells identification and analysis of their status. Method is based on the measurement of the optical density variation in cell suspension after action on it alternative electric field.

Keywords. Cell suspension, electro optical measurement, polarizability, cell status

Introduction

Identification of bacteria cells in the samples is based on the usage of selective labels or selective nutrient mediums. Application of the selective labels provides a fast reply about the presence of sample objects with certain antigenic determinants. These objects can be viable cells, dead cells or their fragments. A drawback of known methods is that the response to fragments of cells will be identified as a false alarm. Application of a selective culture medium allows a determination of viable cells, but does not identify during short time weak, sleeping, dormant and similar kinds of cells. Absence of answers regarding other status of the cells will correspond to their passing. Besides, time of identification for such kinds of cells is expansively long.

The electro-optical method allows combining the advantages of labels application and selective mediums using two step-by-step procedures of measurements. This method is referred to the group of electro physical methods that use the phenomenon of the induced polarization of materials with various properties on a surface of their contact.

Distinction in electric properties of an culture /support medium and a cells, or in electric properties of cells structures, leads to creation of induced charges in an alternative electric field. Charges are created on the border of each pair electrically different mediums [1].

Interaction of induced charges with an electric field leads to occurrence of the rotating moment. Cells change own orientation. They tend to minimal potential energy that corresponds to position of a long axis of a cell in parallel to the vector of an electric field [2]. Change of cell orientation causes change of the optical properties of suspension. Electro-optical measurements are spent at weak orientation of cells, when each cell rotates over a small angle and uniform distribution of cells changes on the small value. However, it is enough, that optical properties of suspension have changed.

Obtaining this variation can be executed at measurement of suspension optical density that can be changed up to 1-2%. For cell identification on a first step change of selective cell surface properties after interaction of its determinants with a selective label – monoclonal or polyclonal antibodies is used [3]. Interaction leads to shielding from an electric field on border the membrane – cytoplasm. As result, the electro-optical signal after them is reduced. For determination of the cells status on the second step their response on the adding of a nutrient medium or medium with power components of growth, for example glucose, is used. Change of cell polarizability and its sizes is investigated simultaneously. With respect to known monitoring methods, that used nutrient mediums, the authentic response to entering of glucose exhibit after 15-30 minutes. It reflects the latent phase of processes that connected with active transport of a substrate inside of cells. In the given work, results of method development for identification of the cells status are presented.

Method

Electro-optical measurements are executed on an EloTrace 2.0 platform from Biotronix GmbH, Deutschland that includes the device, the software and database of electro physical and morphometrical parameters of cells. The device provided continuous performance of the some basic operations with the minimal period 7-8 minutes:

1. Dilution of suspension with unknown optical density and setting fixed parameter N (where N= optical density * volume of suspension);
2. Sedimentation of cells on the filter and their washing by the distilled water;
3. Natural fixation of cells at their removal from the medium and replacement to the measurement medium with low electroconductivity. This procedure results in overlapping by osmotic pressure inside cells of all exchange membrane channels;
4. Resuspension of cells into medium with low electroconductivity g = 5-7 mkSm/sm.
5. Filling by the prepared sample a measuring cell;
6. Performance of electro-optical measurements.

At the electro-optical analysis of cells suspension, stationary value of optical density variation and the form of an electro-optical signal during transition of cells from oriented to chaotic position was measured. The stationary value of optical density variation is proportional to the anisotropic polarizability of cells (AP). This parameter described the absolute value of all induced charges and its distribution. At measurement procedure on frequencies 400-600 kHz this parameter is proportional to the ionic concentration (IK) in the cell cytoplasm or, in other words, is proportional to metabolic activity (MA) of cells. In the given work these three terms AP, IK and MA have identical sense.

The analysis of the relaxation parts of electro-optical signal during cell transition from partial orientation to chaotic position was used for the calculation of the cell mean size. Relaxation of each cell fraction size to chaotic is described by an exponent function.

The exponent index is proportional to the size of cells. The weighting coefficient is proportional to their concentration. For heterogeneous suspension the full signal accumulate sum of signals from homogeneous fractions. Calculation of the square under relaxation curve represents precise estimation of cell mean size.

When results of measurement are analyzed, influence of cells size on the AP value must be taking into account. Registered changes of optical density are induced by change of cell light scattering. For optical approach the abnormal diffraction the light scattering cross section is described approximately by power function of cells mean size with order equal four [4]. Linear dependence of polarizability, as function of mean size, increases in order up to 5 [5].

As model cells cultures E. coli Nissle 1917 and Bacillis Subtilis, that placed before measurements in a various condition were used:

- Active cells E. coli Nissle 1917 after favorable cultivation on the standard growth medium +1% glucose.
- Weaken cells E. coli Nissle 1917 and Bacillis Subtilis after unfavorable cultivation on the standard growth medium +1% glucose, when pH=7.8;
- Cells, after long storage in a dry condition.

Results and discussion

Figure 1 shown results of AP, the mean size of cells and optical density of suspension Ecoli Nisstle as time dependence.

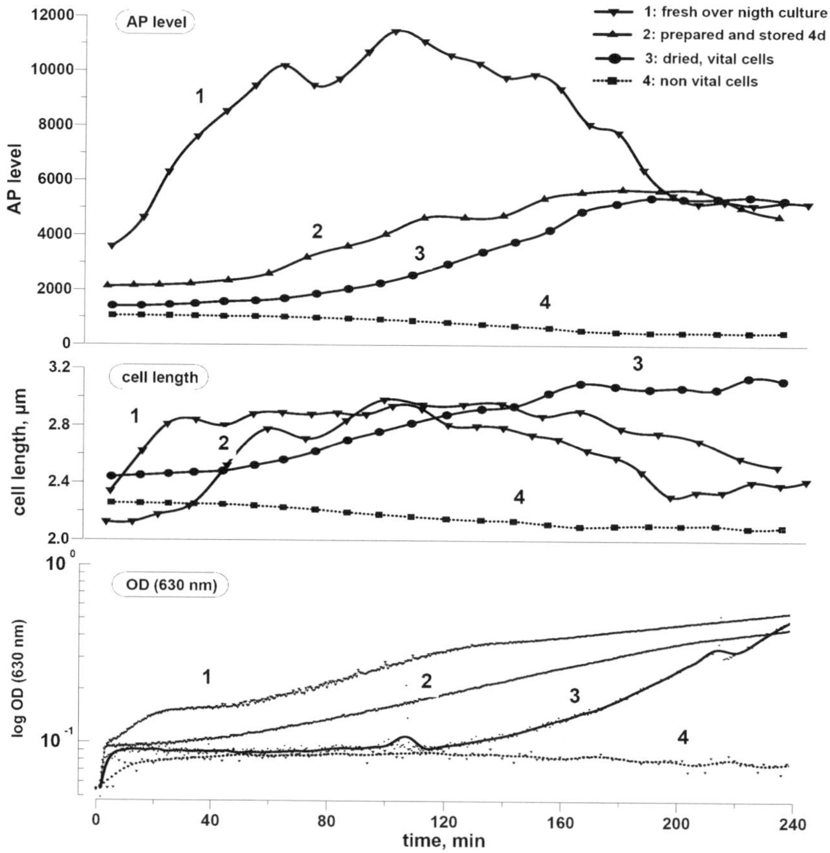

Figure 1. Change of AP, mean size and optical density cell suspension E.coli

Figure 2 shown similar function for cells Bacillis Subtilis.

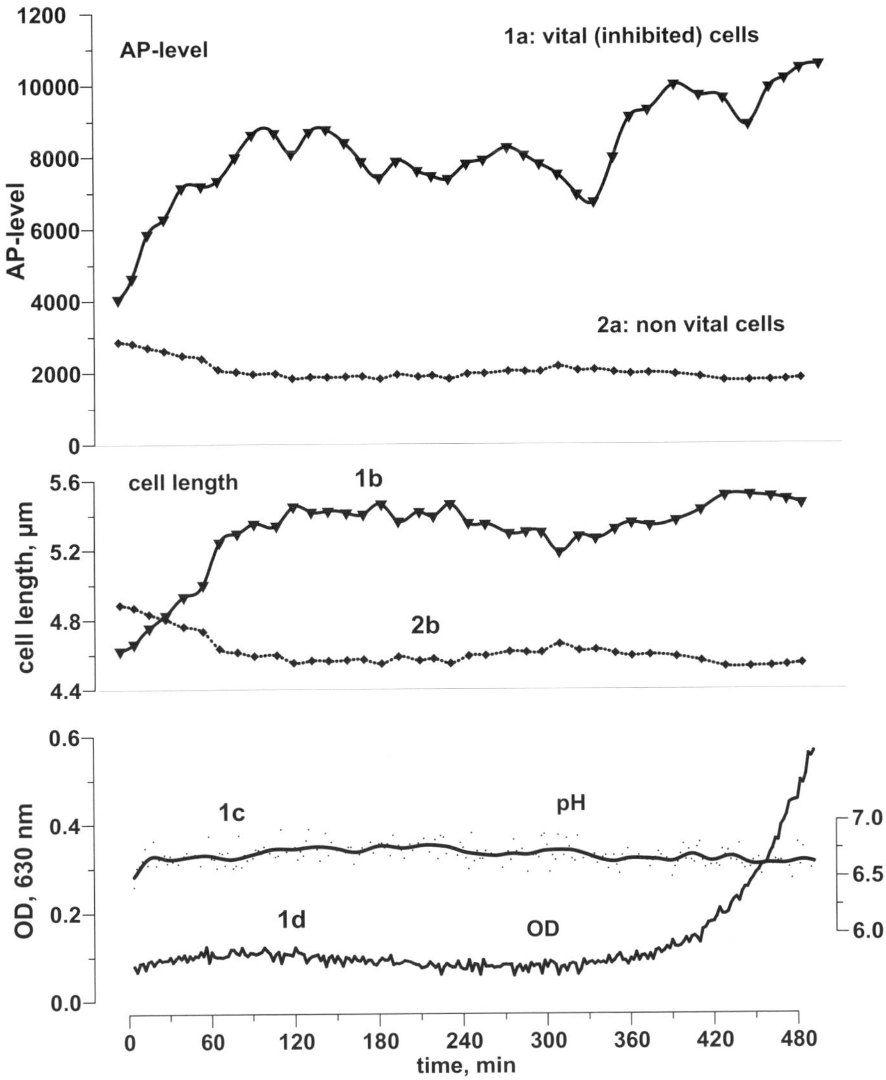

Figure 2. Change of AP, mean size and optical density cell suspension Bacillus subtilis for viable weaken cells after unfavorable growth (pH-7.8) and dead cells.

The represented series of measurements cover possible combinations of real conditions at which the presence of the active, weak, dry and dead cells is probable. It is necessary to note that, traditional microbiological methods for identification cells status by means of optical density or pH variation can not result within scale time 10 - 30 minutes. An exception makes only a case of adding to a nutrient medium of active cells. But this case is not typical for problems of identification and cells status estimation.

The analysis of experimental data allows allocating one feature of non-living bacteria. Cells keep unaffected value of AP. It implied absence of active transport through a membrane inside, and absence of passive diffusion of ions through its outside.

This fact additionally confirms by unchanged mean size of cells. The cell is capable to keep low superfluous osmotic pressure of ions remained in cytoplasm due to rather steady structure of a membrane.

Adding of a nutrient medium influences the active, weak and dry cells identically. But the time scale of changes essentially differs. Immediate active transport of a substrate results in increase in the mean cells size. This process is accompanied by increase in AP of cells. However, after dividing to the mean size function the form of curves changes. During the first 30-40 minutes, normalized AP keeps an unaffected value. In cells occur two processes. Active transport of non dissociated power substrate inside of a cell leads to a decrease of existing ionic concentration and AP value. The decrease should be proportional to the relation of volumes of a cell up to active transport of a substrate and after it. However, it does not occur. Unaffected values of AP reflects stable efficiency of all substrate utilization.

For weakened cells in current of the first 30 minutes, normalized AP is decreased. This fact speaks about absence of conjugate between processes of substrate uptake and its immediate recycling. Only after finishing adaptation process MA of cells is increased. Full time of the ferment systems preparation approximately equal 30-40 mines after beginning of process. However, achievement of a high level of MA, as for active cells, is already impossible. The main part of a power stock of cells was used for restoration of a normal inner status, but not for the growth.

For the dry cells, that are placed in a nutrient medium, normalized AP values do not change. The insignificant increase in the mean size during the first 30 minutes reflects absence of ready system for active transport.

After completion of the transport system preparation the behavior of normalized AP function is similar to weakened cells. For bacterial cells of other kinds measurement of size and AP has similar character. In particular, for cells Bacilus subtilis, as shown in Figure 2, increases of mean size after adding of a nutrient medium to active cells has similar character. In particular, for cells Bacilus subtilis, as shown in **Figure 2**, increases of mean size after adding of a nutrient medium to active cells has similar character. Normalize AP on the used before function of the size led to similar result. Value AP hold unchanged value. Decreasing of the mean size for dead cells of Bacilus subtilis is connected with high level of passive diffusion. Possibly, damage of cells membranes has been expressed in a greater degree. Continuation of an unaffected normalize AP function confirms reliability of used optical and polarizing models.

Conclusion

The electro-optical method of bacterial cell status and identification allows the analysis of their presence in viable or nonviable cell suspensions over a short time. Longer duration electro-optical monitoring of cell growth allows identification of various kinds of viable cells. The given technique provides conclusions about the status of cells, including the case of low concentrations over a short time.

Reference

[1] L.D.Landau, E.M.Liphshitz, Electrodynamics of Continuous medium, Nauka, Moscow, 1982.
[2] S.P. Stoylov, Colloids Electro-optics, Theory, Techniques, Applications, Academic Press, London, 1991.
[3] V.D. Bunin, O.V Ignatov, O.I. Guliy, A.G. Voloshin, L.A. Dykman, D.O'Neil, D.Ivnitski, Studies of Listeria monocytogenes-antibody binding using electro-orientation, Biosens. Bioelectron..19(2004), 1759-1761.
[4] Van de Hulst, Light Scattering by Small Particles, Wiley, NY, 1957.
[5] G.P.F.Bottcher, Theory of electrical polarizability, vol. 1, Academic Press, NY, 1982.

Commercial and Pre-Commercial Cell Detection Technologies for Defence against Bioterror
L.M. Lechuga et al. (Eds.)
IOS Press, 2008

Detection of cells and viruses with mass sensitive devices – applications of synthetic antibodies

A. AFZAL, X. CHEN, M. JENIK, S. KRASSNIG, F. L. DICKERT[1]
Institute of Analytical Chemistry and Food Chemistry, University of Vienna
Waehringer Str. 38, A-1090 Vienna, Austria

Abstract: The label-free, selective and sensitive detection of cells and viruses was successfully performed down to the nanogram and picogram range with acoustic transducers such as the quartz crystal microbalance and the surface acoustic wave resonators. Selectivity of these acoustic devices was optimized by combining them with sensitive layers exhibiting pronounced molecular recognition capabilities based on size, shape and preferably hydrogen bonding. Sensitive layers, often termed as synthetic antibodies, were generated by an innovative method of surface imprinting with bio-analytes. Atomic force microscopy (AFM) proved an excellent tool to examine the bio-imprinted polymer surfaces. Bio-imprinted layers, capable of reversibly absorbing the imprint species, opened up the possibilities to detect different types of cells, for instance, yeast and bacteria. Viruses such as the tobacco mosaic virus, the pox and the human rhinovirus, were specifically detected down to a few ng/mL. Furthermore, by imprinting with bio-analytes, the cross sensitivities can almost be neglected and distinguishing different biogeneous species becomes feasible. The synthetic antibodies yield a more characteristic response pattern than the natural ones.

Keywords: Mass-sensitive devices, Acoustic sensors, QCM, SAW, Synthetic antibodies, Surface imprinting, Bio-analytes.

Introduction

In the recent past, there has been a large flux of scientific research in the fields of bio-technology and genetic engineering, which has revealed a better understanding of the metabolic processes occurring in a living organism and viral infections as well as pathogenic diseases. This research and technology boom has triggered an astonishing development in different areas of bio-science such as medicine, health-care, food technology etc. Moreover, as the public awareness and sensitivity towards hygienic foods and modern-day diseases is increasing with time, new methods of analysis have to be developed in order to avoid threats of harmful biogeneous species such as fungi, viruses, and bacteria. There is an irresistible need to develop fast and straightforward methods for the detection of harmful micro-organisms in contaminated water and food.

[1] Corresponding Author: Franz L. Dickert. Tel.: +43 1 4277 52317; Fax: +43 2243 36273 15; E-mail: Franz.Dickert@univie.ac.at

Although, there are a number of well developed analytical tools designed for the analyses of water and food samples, but they require high-end apparatus and trained personnel and are usually time-consuming. Therefore, massive effort has to be carried out either to miniaturize the analytical appliances or to develop new bio-chemical sensor systems for bio-analyte detection. The later i.e.; the bio-chemical sensors will be discussed in the following review chapter. The chapter deals with the detection of cells and viruses using bio-chemical sensors based upon acoustic devices such as the quartz crystal microbalance (QCM) and the surface acoustic wave (SAW) resonators. Since these devices show a response to mass load (they are also known as the mass-sensitive transducers), and every cell or micro-organism has a certain mass, they can be used for the detection of all types of bio-analytes ranging from biological particles to cells and micro-organisms.

Chemical sensors have already gained considerable importance in the field of analytical chemistry and bio-chemistry, especially when small and rugged systems for remote analyses and/or on-line monitoring [1] of different biological samples are required. Figure 1 gives the general scheme of a chemical mass sensor. In principle, a chemical sensor consists of a chemically sensitive layer, a transducer and the measuring electronics. The sensitive layer interacts with the analyte of interest and hence one of the physical properties of the layer is changed (e.g.; mass, in this case). Change in the mass of the sensitive layer is directly sensed by the mass-sensitive transducer i.e.; either a QCM or a SAW, due to a relative change in its resonance frequency. The transducer then transforms these changes to an electrically measurable quantity.

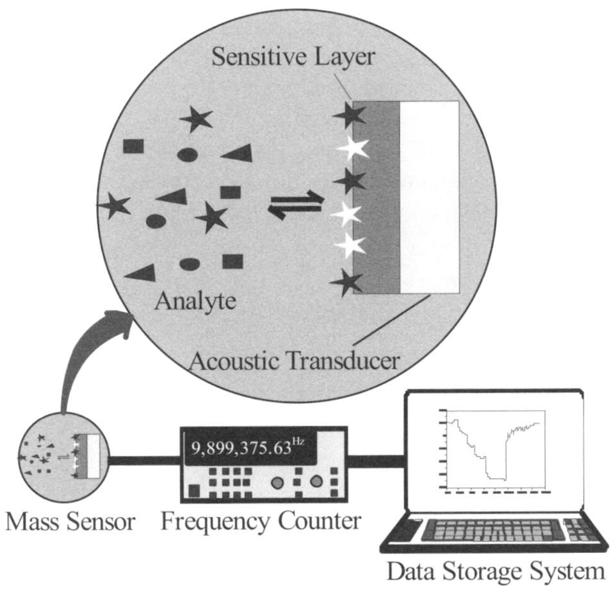

Figure 1: Principle of a chemical sensor based on an acoustic transducer.

1. Mass-Sensitive Devices

Mass-sensitive devices such as quartz crystal microbalance (QCM) and surface acoustic wave (SAW) resonators are principally derived from the piezoelectric effect. According to Jacques Curie and Pierre Curie, who first described the piezoelectric effect in 1880, when a mono-crystalline piezoelectric material is subjected to a physical stress, a measurable voltage is produced on the crystal surfaces [2]. Naturally, the reverse of this effect, the so-called inverse piezoelectric effect, is also possible i.e. on applying an alternating current to a piezoelectric material, a mechanical oscillation can be induced in the crystal. Various piezoelectric materials such as quartz, lithium tantalate and lithium niobate can be applied expediently in chemical sensing. However, quartz is the most important substance in this field because of the low dependence of its oscillation frequency on temperature.

1.1. The Bulk Acoustic Wave Devices

The nature of oscillation (i.e. the bulk or the surface oscillation) induced in the device on applying a certain voltage and the temperature behavior of the crystal depends upon the cutting angle of the device substrate with respect to the crystallographic main axis. The most extensively used mass-sensitive device is QCM consisting of an AT-cut (35° 15′) quartz plate on which a desired electrode structure is applied on both sides (see Figure 2). When an electrical potential is applied to the QCM, a thickness shear wave is generated i.e.; the two faces of the quartz move apart from each other, as shown in Figure 3. The elementary resonance frequency of such an oscillating crystal depends on the thickness of the quartz plate. Commercially QCMs are available with resonance frequencies up to 50 MHz. At higher frequencies, the quartz becomes too thin to be mechanically stable, for instance, thickness of 20 MHz quartz is 84μm only. However, the higher frequencies can be reached in two ways [3]; either by operating the quartz at a resonance of overtones or by partially etching the QCM.

In 1959, it was shown by Sauerbrey [4] that the resonance frequency of an oscillating quartz crystal is dependent on the mass applied to the electrodes (see also Figure 3). Equation 1 gives the mathematical expression;

$$\Delta f = -\frac{1}{\rho_m k_f} f_o^2 \frac{\Delta m}{A_{cr}} \qquad \text{.................... (1)}$$

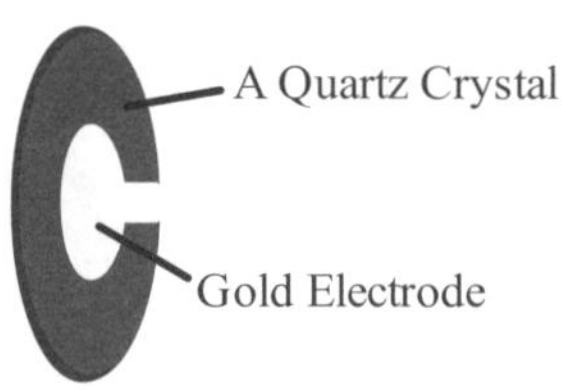

Figure 2: A quartz crystal microbalance.

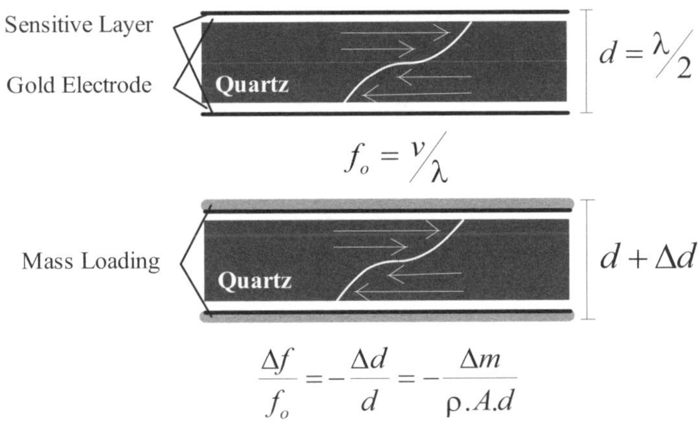

Figure 3: The principle of a QCM; showing generation of a thickness shear wave and the effect of mass load on the resonance frequency of a quartz.

The Sauerbrey equation shows that the change in resonance frequency, Δf, caused by a mass change on the electrodes depends directly on the deposited mass and on the square of the fundamental frequency. The Sauerbrey relationship is generally followed in gas phase measurements. For the liquid phase measurements, however, the properties of the liquids have to be considered. In 1985, Kanazawa and Gordon [5] derived a relationship between change in frequency of a quartz and the properties of a liquid i.e.; viscosity and density. Equation 2 gives the mathematical expression of the frequency shift while placing the crystal into a liquid environment.

$$\Delta f = -f_o^{3/2}\left(\frac{\eta_l \rho_l}{\pi \eta_q \rho_q}\right)^{1/2} \qquad \text{.................... (2)}$$

The equation relates the change in frequency to the density and viscosity of the surrounding liquid as well as to the density and viscosity of the quartz crystal. Theoretically expected frequency shift from air to water or aqueous solution for a 10 MHz crystal is about 2 to 4 kHz. However, this value is usually underestimated. The observed frequency shift is often 1.5 to 3 times higher. This discrepancy is attributed to the differences between the surface and bulk values of viscosity and density, hydrophilic or hydrophobic nature of the surface, intermolecular forces, electric double layer structure, etc. If the solution is in contact with both oscillator faces, then the frequency change is doubled.

The Sauerbrey equation also explains why the QCMs are regarded as highly sensitive microbalances, because a 10MHz QCM has a sensitivity of 1Hz per nanogram of the mass loading. The sensitivity of the devices can further be enhanced by using QCMs with higher intrinsic resonance frequencies. The greater the fundamental resonance frequency of quartz, the higher is the sensitivity of the device. This can be seen in Figure 4, where the sensor effect of different QCMs is plotted as function of their fundamental frequencies. These QCMs were exposed to polyvinyl chloride (PVC) solution in tetrahydrofurane (THF) at 25°C.

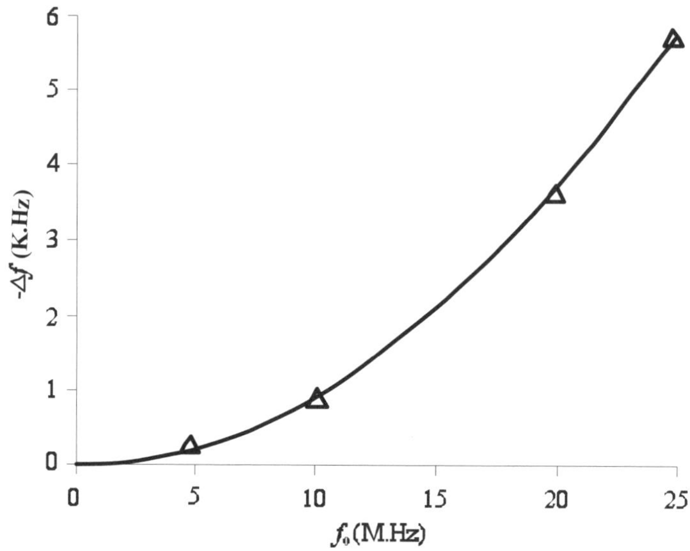

Figure 4: Sensor response of different QCMs to low molecular weight Polyvinyl chloride (PVC) in Tetrahydro Furan (THF) plotted as a function of the elementary resonance frequency of the respective QCMs.

In gas phase, the deviations of the actual system from the ideal one are usually small and insignificant. In liquid phase, however, the QCM shows substantial response to temperature changes, principally because the sample viscosity is highly temperature dependent. It can be observed in Figure 5, where the frequency signals of a dual-electrode 10MHz QCM immersed in water are shown for 1°C rise in temperature from 20-21°C. These results have already been published by our group [6]. Evidently, one degree rise in temperature results in a frequency shift of 50Hz, which is not a negligible value, especially when very low concentrations of the analyte are to be detected.

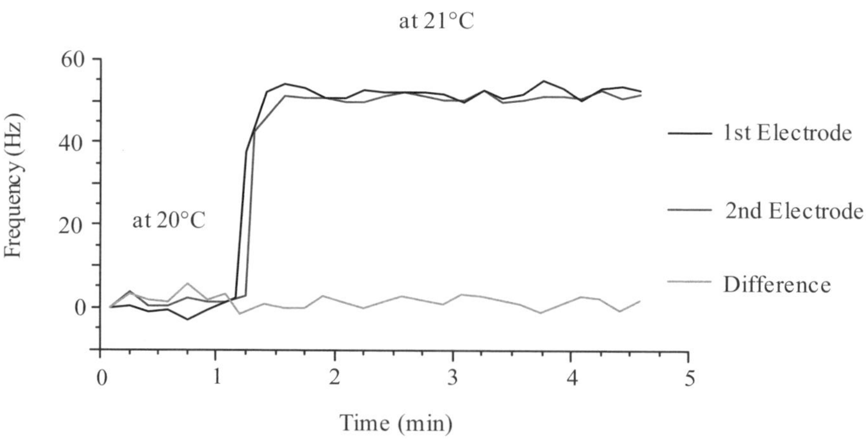

Figure 5: Sensor response for 1°C rise in temperature of a two-electrode QCM immersed into water with one face.

The surrounding liquid molecules interact with the QCM and due to adhesion, they will be accelerated. Consequently, the resonance propagates into the surrounding medium and the quartz surface oscillation is damped due to friction, which causes a frequency shift and a change in electronic properties. However, as depicted in Figure 5, the difference between the two channels remains constant. Hence, the total compensation of the temperature and the physical effects can be made by applying a dual-electrode structure on the QCM, coating one of the electrodes with a sensitive layer and using the second as a reference.

1.2. The Surface Acoustic Wave Devices

In principle, the surface acoustic wave (SAW) devices consist of an ST-cut quartz plate, on which two interdigital (finger) electrodes are deposited (see in Figure 6). The application of an alternating current creates an electric field in the piezoelectric material and each point within the crystal is displaced. As a result, a Rayleigh wave is produced, as shown in Figure 7(A), which subsequently travels along the surface of the sensor until it interacts with the other electrode. This triggers an alternating voltage, which can then be detected electrically and quantified. The resonance frequency of a SAW device depends upon the dimensions of the electrode structure, because the wavelength of the propagating wave is determined by the distance between the centers of two equally polarized finger electrodes. SAWs can be operated in the frequency ranges up to 2.5 GHz and analytical applications have been reported up to 1.0 GHz [7]. This shows significantly improved sensitivity of the SAW devices towards mass loading compared with QCMs, for instance, with a 434 MHz SAW, absolute mass changes as low as one picogram are detectable.

The drawback of the SAW is that it can not be used in the liquid phase owing to the extreme damping of the surface wave caused by the higher sensitivity of the device towards viscosity and mass loading. However, this can be overcome with an alternative design that is the shear-transverse wave (STW) resonators [8, 9]. In contrast to SAW, such devices consist of lithium tantalate as transducer material. If a potential is applied to the STW, it results in a surface transverse wave propagating exclusively within the substrate surface, see also Figure 7(B). In this case, there are no or only negligible viscous interactions with the surrounding medium, so the primary damping occurring in SAW can be avoided. Similarly to QCMs, surface acoustic devices also measure the difference between the actual sensing channel and a reference to compensate temperature and viscosity effects.

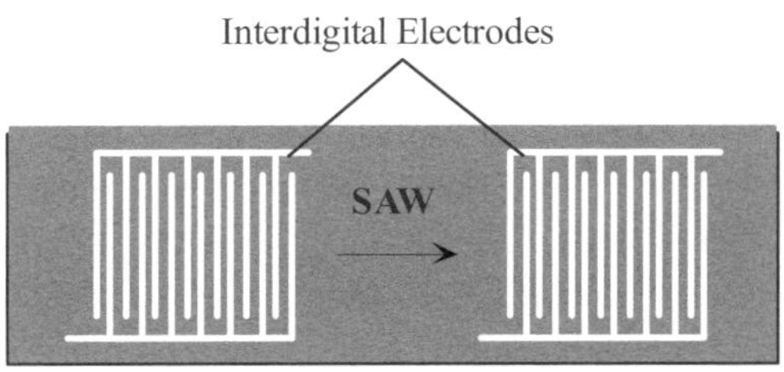

Figure 6: A surface acoustic wave device.

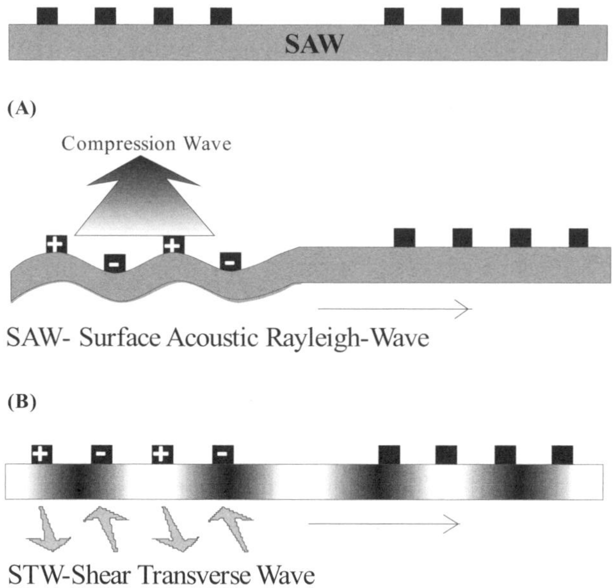

Figure 7: Principle of the surface acoustic wave devices: showing the generation of (A) a surface acoustic Rayleigh-wave, and (B) a shear transverse wave.

2. Sensitive Layers

To develop a bio-chemical sensor system for a distinct analytical problem, it is essential to generate selectivity towards a specific analyte so as to distinguish between different types of cells. The mass-sensitive transducers alone cannot discriminate between different analytes: either different types of biological cells or viruses, and hence they need a sensitive layer. Piezoelectric crystals can be coated by a chemical recognition layer to provide selectivity to the sensor. In some cases, the metallic electrode applied on the crystal can itself be the selective film, for instance, the gold electrode can directly be applied to detect thiols. However, considering biological analytes, mostly these devices are coated with a dye showing a response to the surrounding biological environment. In general, there are two types of chemically sensitive layers [10]: one that interacts with the analyte chemically and forms a covalent bond, and the second offers spatially fitting cavities for analyte inclusion based upon purely non-covalent interactions, for example, hydrogen bonds, π-π interactions and/or dipolar interactions. Currently, different layer design strategies are used for the detection of bio-analytes, such as enzyme-substrate binding, ligand-receptor binding, immunological recognition and artificial polymers or synthetic antibodies.

2.1. Synthetic Antibodies—Imprinting

It is rational to apply biological recognition materials to the bio-chemical sensors. However, in spite of the outstanding sensitivity and selectivity of the biological recognition layers, they have the intrinsic disadvantage of comparably low long-term stability. Biological compounds usually degrade or denaturize, which strongly reduces the possible measurement cycles as well as the shelf-life of the respective sensor. An elegant layer design strategy to overcome these limitations as well as to generate cavities ideally adapted to the analyte-to-be is to imprint a highly cross-linked polymer matrix with the desired template. The method combines a straightforward synthetic procedure with an exceptional variability and flexibility in application [11]. Molecular imprinting strategies were primarily developed by Wulff [12] and Mosbach [13, 14]. Recently, these methods have become increasingly accepted for the design of artificial antibodies against a variety of analytes.

In molecular imprinting, the template is mixed with the respective monomers and cross-linkers, subsequently followed by the polymerization reaction. After completion the template is either removed by washing or by heating (evaporation) leaving behind interaction sites with in the whole polymer bulk. Small analyte molecules can then diffuse through the pores and re-incorporate into the polymer layer. This strategy leads to remarkable sensitivity and selectivity of the chemical sensors in order to detect small molecular analytes like nandrolone [15] and xylene isomers [16]. Certainly, this technique is not limited only to molecules but is also applicable to larger species i.e.; bio-analytes such as cells and viruses. One of the first attempts was successfully made by Alexander and Vulfson [17] by imprinting micro-beads and polymers with bacteria.

2.2. Surface Imprinting with Bio-analytes

A primary difference between imprinting with a biological species and a molecular compound is that in case of bio-imprinting bulk imprinting procedures can not be applied due to the larger dimensions. The larger dimension of the bio-analytes substantially hinders the diffusion in the polymer bulk. Therefore, surface imprinting is much more practical in this case [18]. Figure 8 depicts the stamping strategy for imprinting biological species on the polymer surface. Here, the template species is self-assembled on a flat surface to form a stamp. In parallel, a pre-polymerized oligomeric mixture is deposited on the transducer. The stamp is then pressed into the polymerizing material; as a result the polymer chains organize themselves around the bio-template during hardening. Finally, the template is removed from the polymerized material leaving behind the mould, so the resulting layer can be regarded as the synthetic antibody analogue. The selectivity of such a layer depends on size and shape of the template as well as on the respective surface chemistry

Surface imprinted synthetic antibodies have already been prepared by polymerizing a monomer solution to its gel-point. Polymers synthesized either by radical polymerization or by polyaddition reactions can be used for surface imprinting [19]. The exemplary results of this process have already been presented [20], which were obtained with polyurethane layers composed of bisphenol A and 4, 4′-diisocyanato diphenylmethane. Phloroglucinol served as the cross-linker. These were mixed and pre-polymerized at 70°C until the gel-point was reached. The gel-like prepolymer was diluted with tetrahydrofurane (THF) to an appropriate concentration. Thin films of the pre-polymer were formed either by spin coating or by drop coating the respective electrodes on a transducer. Surface imprinting was performed with the help of a stamp of purified templates pressed onto the polymerizing thin film during the curing. The stamp was then removed by a non-destructive lift-off process so that the sensitive layer did not peel off. The polymer surface was rinsed with water or sodium dodecyl sulphate (SDS) in an ultrasonic bath to remove the templates.

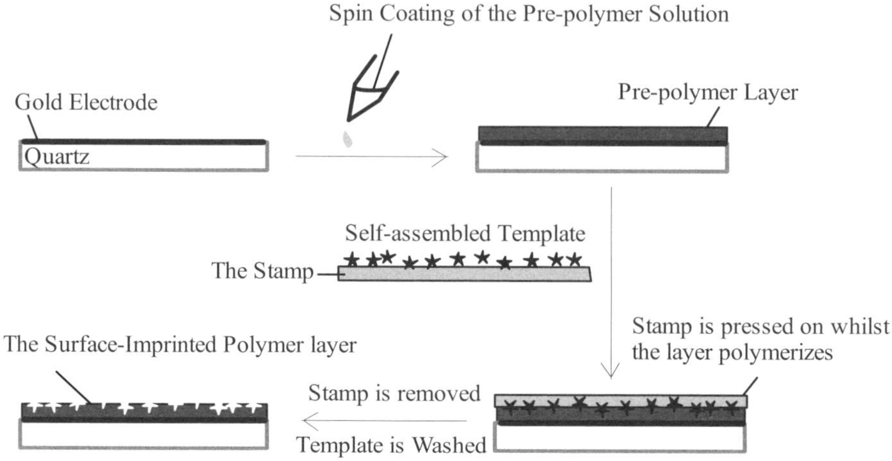

Figure 8: Schematic representation of the surface imprinting technique for bio-analytes; the imprint material is self assembled on a stamp.

3. Sensor Applications

3.1. Yeast Cells

Yeasts are eukaryotic micro-organisms, classified as the prevalent members of the kingdom fungi, with approximately 1500 species [21]. Some species of yeast are pathogenic and they can cause irritation, shedding of the tissues and oral or vaginal infections in human. So there is a need for the detection of yeast cells, which was accomplished by surface imprinting polyurethanes with *Saccharomyces cerevisiae* i.e.; a harmless species of yeast, vastly used in baking and fermentation processes. Figure 9 shows the AFM-image of a polyurethane layer imprinted with yeast cells [22]. It can be observed that each cell casts its own impression in the material after the templates are washed out. The resulting cavities are hexagonal and their dimensions vary according to the size distribution of the cells. The yeast cells adopt the most compactly packed surface structure i.e.; honeycomb-like packing, during stamp production. When such a polymer is exposed to a solution containing yeast, the cells adhere to its surface due to the sterically adapted cavities.

The significance and application of the surface imprinting technique can be estimated via AFM so that even some biological processes can be imaged within the material. Yeast cells usually reproduce asexually by means of budding [21]. Therefore, a yeast sample contains cells at different stages of growth, the dimensions of which are readily reproduced into the imprinted layer. Figure 10 shows an inverted image of the yeast-sensitive polymer layer, where the daughter cells (or buds) are clearly distinguishable.

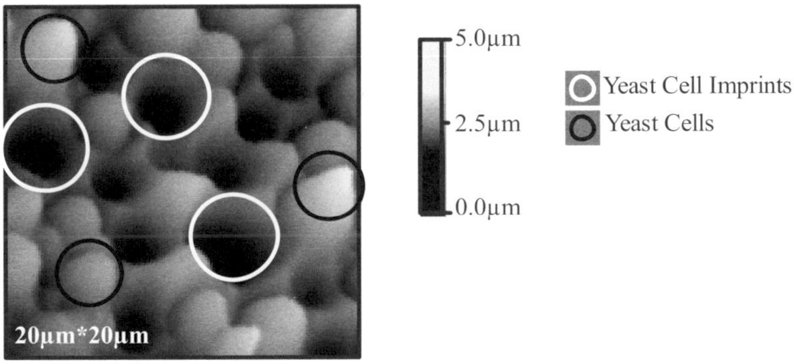

Figure 9: AFM image of the surface of a yeast-imprinted polyurethane layer. Some of the cavities are re-occupied by the yeast cells from the surrounding media.

Yeast-imprinted polyurethane layer was applied on a dual channel piezo-crystal alongside a non-imprinted reference electrode to perform mass-sensitive measurement in the aqueous phase. The measurement was performed at 25°C with a flow rate of 5mL/min. The *S. Cerevisiae* cells were suspended in a 1/15 molar KH_2PO_4/Na_2HPO_4 buffer of pH 6. Figure 11 shows the overall mass effect obtained by the absorption of yeast cells into the layer material. The frequency change here is the difference of the sensitive electrode and the reference electrode. The non-specific adhesion of the yeast cells was insignificant. In contrast, the imprinted polymer layer shows excellent adhesive properties for yeast cells. The cells incorporated into the cavities were washed off with a short pulse of 100mL/min. It was evident that the interactions are completely reversible. Hence, chemical sensors based on imprinted polymers can be used for online label-free detection of micro-organisms without using fragile antibodies as in case of piezoelectric bio-sensors [23]. Therefore, limitations such as regeneration of sensitive layers, washing and drying steps as in offline piezoelectric measurements, and long incubation times are overcome. Moreover, the imprinted polymer layers can be optimized to selective detection of different species of yeasts.

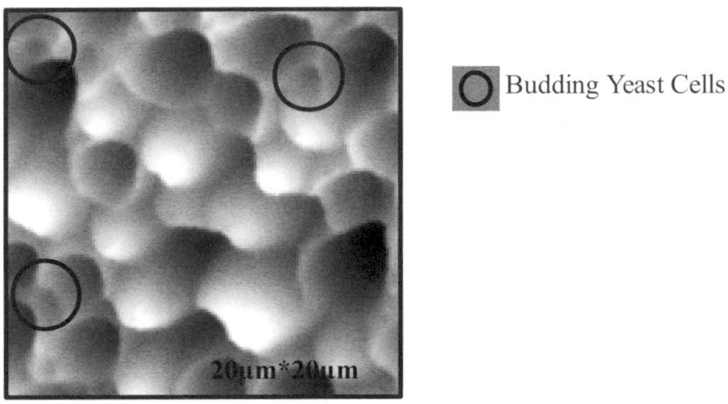

Figure 10: Inverted image of a yeast-imprinted polyurethane layer showing budding process.

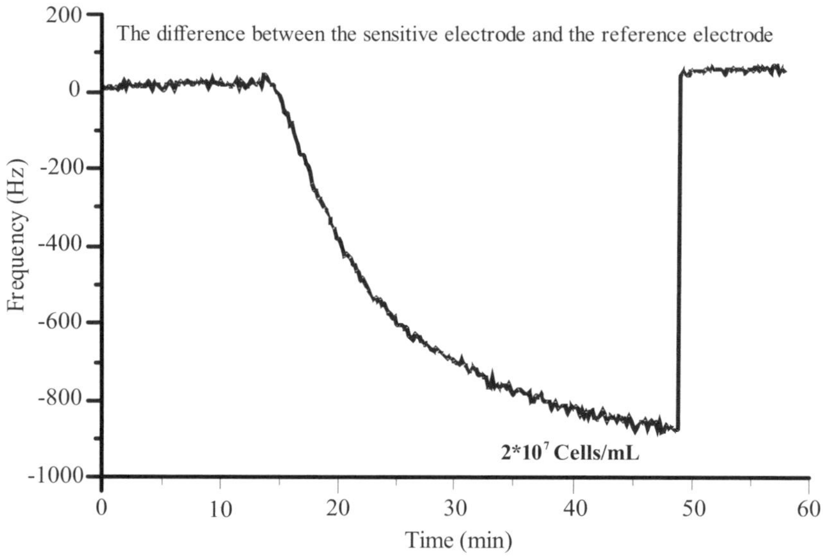

Figure 11: Sensor effect of a 1mg/mL *S. Cerevisiae* in phosphate buffer of pH 6 at 25°C. The non-specific adhesion at the reference electrode was subtracted to get the overall mass-effect.

3.2. Bacteria

Bacteria are the unicellular micro-organisms with varying sizes and shapes. Though, certain species of bacteria are useful, some are really harmful. The detection of such harmful species is a key task. Some basic results have already been published [24]. Here, we present results obtained from the mass-sensitive measurements using surface imprinted polyurethane layers coated on the QCM. Surface imprinting provides yet another example of the detection of biogeneous analytes in liquid phase. *Escherichia coli*, present in the lower intestine of mammals, was studied as an experimental model.

Figure 12 shows the AFM image of a polyurethane layer stamped with *E. coli* during polymerization. The imprinted surface shows prominent cavities within the layer. Some of these cavities are occupied by bacteria illustrating a stronger adherence of the imprints to the surface. Humidity content, pre-polymerization time and curing conditions collectively play an important role in obtaining optimal interactions between the layer and the imprint material [25]. Due to much smaller dimensions of the bacterial cells, the pits and walls are not as pronounced as in case of yeast cells. Nevertheless, the layer offers sterically adapted interaction centers for the re-inclusion of bacteria.

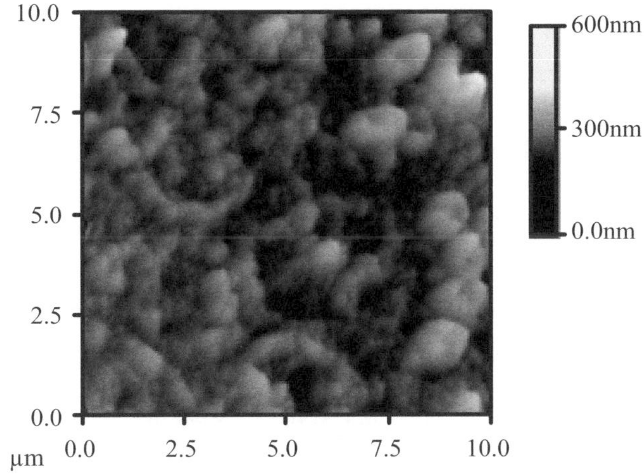

Figure 12: Surface-structured polyurethane layer with *E. coli* imprints.

The mass-sensitive sensor system was established by coating a two-electrode QCM with the imprinted and non-imprinted polyurethane layers. The QCM was then exposed to a suspension of *E. coli* containing 1.0mg/mL bacterial cells at 25°C. A phosphate buffer of pH 6 was used for making the suspension. Figure 13 gives the sensor response of the two electrodes to different concentrations of E. coli cells. The frequency decrease of the imprinted channel clearly indicates cell deposition in the imprints, which provide binding sites for the appropriate adhesion of bacteria. In contrast to this, the reference channel does not show any response to the micro-organism. Although the chemical composition of the layers was same, the imprinted layer shows a much higher sensor effect as compared to the reference. Actually, the non-imprinted polymer does not offer any kind of interaction sites for the detection of bacteria.

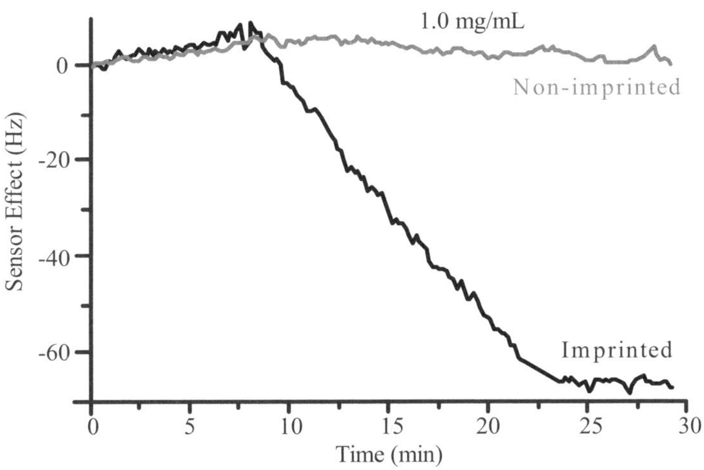

Figure `13: QCM sensor response for *E. coli* imprinted and non-imprinted polyurethane layers to different suspensions of *E. coli* in a phosphate buffer (pH 6) at 25°C.

3.3. Viruses

Viruses are the biological particles that can infect plants and animals, so they are the next to be detected with the imprinted sensor layers. They have smaller size in the range of 20-300nm and hence cannot be detected by microscopes; therefore no straightforward method exists for their detection. However, acoustic sensors coated with synthetic antibodies can do the job. In the following section, different types of viruses would be discussed as exemplary sensor applications.

3.3.1. Tobacco Mosaic Viruses (TMV)

The tobacco mosaic virus (TMV) is an RNA virus that infects tobacco and other plants of the family *Solanaceae*. It consists of an inner helical RNA strand covered with a protein case. It has a cylindrical shape with a length of 300nm. The external and internal diameters are 18 and 4nm, respectively. These viruses tend to form hexameric aggregates at higher concentrations, where six rods align themselves along each other [18]. TMV imprinted polymer layer was prepared by surface grafting a monolayer of the viruses on a stamp, which was pressed into the polymerizing material. In this case, styrene-methacrylate copolymer was used. After polymerization, the stamp was removed and the layer was washed with 0.2% (w/w) aqueous solution of sodium dodecyl sulphate (SDS) leaving behind the adapted cavities. In the AFM image (see Figure 14), the resulting pits can be seen clearly, which evidently reproduce the steric features of the TMV. According to Bittener [26], TMV interacts with the surface of the polymer due to different functional groups present within the protein shell. Therefore, to avoid covalent bonding of the viruses to the polymerizing layer, a protective layer of amylose was used to cover the stamp before applying.

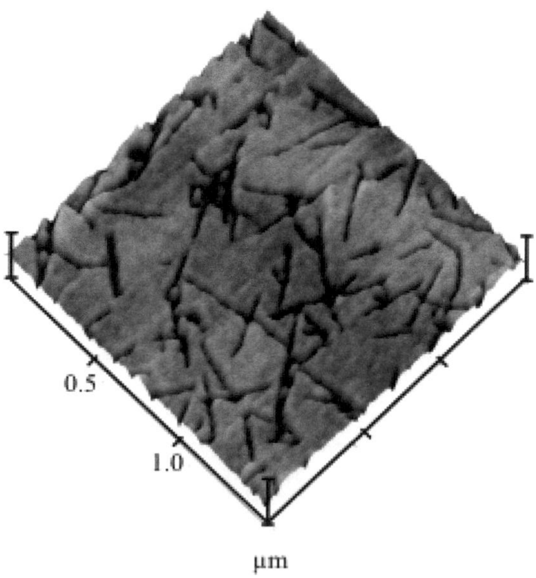

Figure 14: AFM image of a TMV imprinted polymer layer.

3.3.2. Human Rhinovirus (HRV)

Human rhinovirus serotype-2 (HRV-2) is icosahedrally shaped with a diameter of 30nm and is the cause of common cold in human beings. The virus moulds were prepared using surface imprinting techniques and coated onto a QCM to perform the mass-sensitive measurements.

3.3.2.1. Selectivity

The surface imprinting process leads to perfect cast of viruses, which not only makes the layer highly sensitive but also the selectivity is induced and different types of viruses can be clearly distinguished. Cross-sensitivity of the HRV-imprinted layer was observed by mass-sensitive measurements with HRV-2 as well as with TMV. Figure 15 shows the QCM-sensor response for two different experiments. When a HRV-imprint was exposed to a solution of $7.2*10^{10}$ particles/mL HRV-2, the frequency dropped around 1300Hz. While with the same imprint, TMV could not be detected at all. It is evident that the surfaces are strongly modified by imprinting. Hence, these results indicate the size and shape recognition capabilities of the surface grafted polymer materials.

3.3.3. Parapox Ovis Viruses

Bio-imprinting strategies are also applicable to much more fragile species than rather robust TMV. Parapox ovis virus (PPOV) is a relatively harmless model species for smallpox. PPOV consists of two cores. The inner core contains the actual genetic information that is surrounded by a protein palisade layer. The outer core contains two liquid-filled lateral bodies. These characteristic features of a PPOV make it sensitive to mechanical stresses [18]. Therefore, the stamping procedure was modified and the materials from soft-lithographic techniques [16, 27] such as PDMS (polydimethyl-siloxane), a soft silicone rubber, as the stamp support. Figure 16 shows the AFM image of PPOV-imprinted polyurethane, where the interaction sites can be identified clearly in the polymeric material.

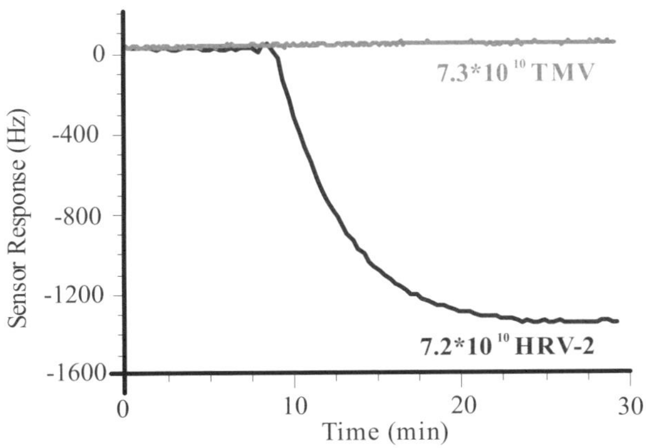

Figure 15: Cross-sensitivity of HRV-2 imprinted polymer layer; the frequency shifts are corrected by the reference signal.

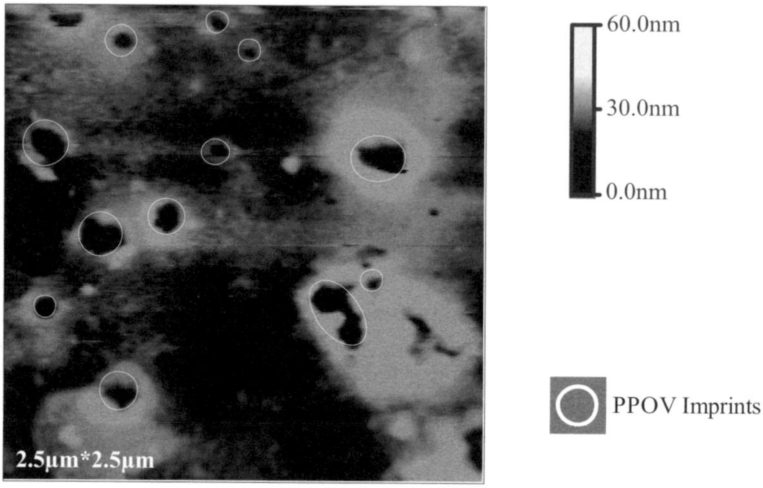

Figure 16: AFM micrographs showing a polyurethane surface imprinted with the parapoxovis virus (PPOV).

Coating a QCM with this layer material leads to a PPOV-sensor. The response of a dual-electrode QCM towards PPOV suspension in water is given in Figure 17. One of the channels was coated with a non-imprinted material and acts as reference. It is evident that the imprinted layer takes up much more virus particles as compared to the non-imprinted layer. The effect due to pure sedimentation of the virus on the surface of the layer can be excluded by taking the difference of the two layers.

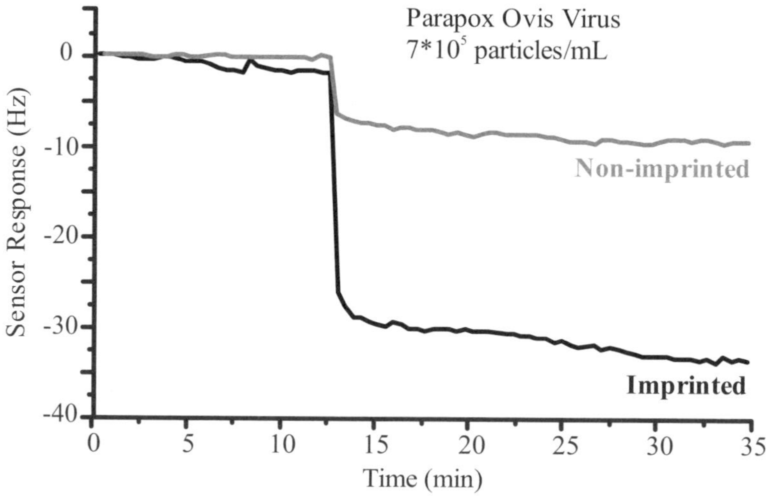

Figure 17: Sensor response of a dual-electrode QCM to PPOV suspensions. The sensitive electrode was coated with PPOV-imprinted polyurethane, and the reference with a non-imprinted polymer.

Conclusion

Surface imprinting with bio-analytes to generate synthetic antibodies for the detection of harmful biological species has undoubtedly proven an exceptional tool for sensor design. Surface imprinting strategy offers a straightforward way to design highly selective polymer layers with antibody-like interaction properties. Different types of bio-analytes, such as bacteria, viruses and yeasts, can be successfully detected to nano-gram levels. The work presented here deals with rather harmless biogenous species to avoid preliminary risks. However, this strategy can be extended to other more dangerous species as it has demonstrated a lot of promise and confidence towards sensitive detection of bio-analytes. Moreover, the selectivity of the sensor layers can be tuned to differences in size and shape of the imprint material. The resulting polymers are able to selectively re-incorporate the analyte of interest. Combining them with the mass-sensitive devices, such as quartz crystal microbalances, leads to a selective, concentration dependent sensor response. The technique combines the specificity of bio-recognition with thermal, mechanical and chemical stability of the synthetic antibodies to allow fast, reliable online measurements, which was not yet possible.

References

[1] Franz L. Dickert, Peter A. Lieberzeit, *Encyclopaedia of Analytical Chemistry, Meyers RA (edn.), John Wiley & Sons, Chichester*, 2000: 3831-3855
[2] A. Janshoff, H. J. Galla, C. Steinem, *Angew. Chem. Int. Ed.* 2000: 39, 4004-4032
[3] S. Buettgenbach, J. Rabe, B. Zimmermann, P. R. Hauptmann, *Proc. SPIE*, 2001: 4205, 207
[4] G. Sauerbrey, *Zeitshrift für Physik*, 1959: 155, 206-222
[5] K.K. Kanazawa, J.G. Gordon, *Analytical Chemistry*, 1985: 57, 1770-1771
[6] P. A. Lieberzeit, G. Glanznig, M. Jenik, S. Gazda-Miarecka, F. L. Dickert, A. Leidl, *Sensors*, 2005: 5, 509-518
[7] F. L. Dickert, P. Forth, W. E. Bulst, G. Fischerauer, U. Knauer, *Sensors and Actuators B*, 1998: 46, 120–125
[8] J. B. Briot, S. Ballandras, E. Bigler, G. Martin, *Proceedings of 1997 IEEE International Frequency Control Symposium, Orlando, Florida, IEEE Cat. No. 97CH36016, IEEE,* Piscataway, NJ, 1997: 207–212
[9] M. Tom-Moy, R. L. Baer, D. Spira-Solomon, T. P. Doherty, *Analytical Chemistry*, 1995: 67, 1510–1516
[10] Franz L. Dickert, Peter A. Lieberzeit, *Springer Series on Chemical Sensors and Biosensors, Springer-Verlag, Berlin Heidelberg*, 2006: 5
[11] Sellergren, *Molecularly imprinted polymers-techniques and instrumentation in analytical chemistry, 1st edn. Elsevier, Amsterdam*, 2001: p. 23
[12] G. Wulff, *Angew. Chem. Int. Ed. Engl.* 1995: 34, 1812
[13] A. G. Mayes, K. Mosbach, *Trends Anal. Chem.* 1997: 16, 321
[14] K. Mosbach, K. Haupt, *J. Mol. Recog.* 1998: 11, 62
[15] C. J. Percival, S. Stanley, A. Braithwaite, M. I. Newton, G. McHale, *Analyst*, 2002: 127, 1024
[16] P. A. Lieberzeit, S. Gazda-Miarecka, K. Halikias, C. Schirk, J. Kauling, F. L. Dickert, *Sensors and Actuators B*, 2005: 111-112, 259
[17] C. Alexander, E. N. Vulfson, *Adv. Mater.* 1997: 9, 751
[18] Oliver Hayden, Peter A. Lieberzeit, Dieter Blass, Franz L. Dickert, *Adv. Funct. Mater.* 2006: 16, 1269
[19] B. Ebarvia, F. Sevilla, *Sensors and Actuators B*, 2005: 107, 782
[20] O. Hayden, D. Podlipna, X. Chen, S. Krassnig, A. Leidl, F. L. Dickert, *Mater. Sci. Engin. C*, 2006: 26, 924-928
[21] C. P. Kurtzman, J. W. Fell, *Yeast systematics and phylogeny - implications of molecular identification methods for studies in ecology.* In: C. A. Rosa, G. Peter, editors; *The Yeast Handbook. Germany: Springer-Verlag, Berlin Heidelberg*, 2006: p. 11-30
[22] F. L. Dickert, O. Hayden, P. A. Lieberzeit, C. Haderspoeck, R. Bindeus, C. Palfinger, B. Wirl, *Synthetic Metals*, 2003: 138, 65-69

[23] D. Ivnitski, I. Abdel-Hamid, P. Atanasov, E. Wilkins, *Biosens. Bioelectron.* 1999: 14, 599
[24] Oliver Hayden, Franz L. Dickert, *Adv. Mater.* 2001: 13, 19, 1480
[25] Franz L. Dickert, P. A. Lieberzeit, O. Hayden, R. Bindeus, K-J. Mann, C. Haderspöck, *Mat. Res. Soc. Symp. Proc.* 2003: 776
[26] M. Kenz, M. Sumser, A. M. Bittner, C. Wege, H. Jeske, D. M. P. Hoffmann, K. Kuhnke, K. Kern, *Langmuir*, 2004: 20, 441
[27] R. S. Kane, S. Takayama, E. Ostuni, D. E. Ingber, G. M. Whitesides, *Biomaterials*, 1999: 20, 2363

Commercial and Pre-Commercial Cell Detection Technologies for Defence against Bioterror
L.M. Lechuga et al. (Eds.)
IOS Press, 2008

Cell monitoring systems with CMOS micro-sensor-chips

Werner BAUMANN[a], Carsten TAUTORAT[a], Angela PODSSUN[a], Philipp KÖSTER[a], Jan GIMSA[a]; Ralf EHRET[b], Ingo FREUND[c], Mirko LEHMANN[c]
[a] *University Rostock, Biophysics Institute, Gertrudenstr. 11A, 18057 Rostock, Germany*
[b] *Bionas GmbH, Friedrich-Barnewitz-Str. 3, 18119 Rostock Germany*
[c] *Micronas GmbH, Hans-Bunte-Str. 19, 79108 Freiburg, Germany.*

Abstract: A better understanding of the multifunctional cellular processing of input- and output-signals in living cells is fundamental for basic research, development of drugs and for environmental monitoring e.g. the detection of bio-toxic agents. For on-line monitoring of cellular reactions we develop(ed) several Cell Monitoring Systems (CMS®). They allow the parallel and non-invasive measurement of different parameters of living cells by the use of CMOS silicon microsensors.

Keywords: biosensor, CMOS microchip, environmental monitoring, neuronal network, CMS, Cell Monitoring Systems

Introduction

Biochemical substances are sensitively recognized and processed by living cells, either to provide life-energy or to trigger an adequate cell-type specific response. A better understanding of the multifunctional cellular processing of input- and output-signals is fundamental for basic research as well as for various fields of biomedical applications and for environmental monitoring e.g. the detection of bioterror agents. As a first approach for on-line monitoring of cellular reactions under well controlled experimental conditions we develop(ed) several Cell Monitoring Systems (CMS®). They allow the parallel and non-invasive measurement of

Figure 1: CMOS sensorchip with encapsulation on standard 68-pin PLCC ceramic socket. Cells are cultured directly in the well on chip.

different parameters of cellular systems by the use of CMOS silicon microsensors, realized in cooperation with the semiconductor company Micronas [1-6]. At present the main application for our system is in the field of high content screening in the drug development process. In environmental monitoring applications this system allows the detection of bio-toxic agents by using living cells as detection targets.

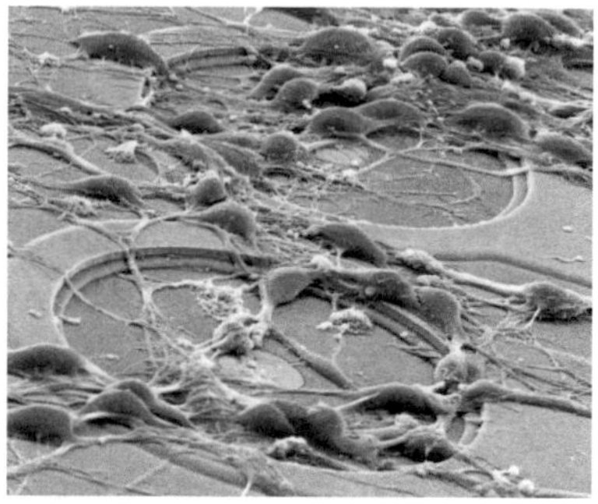

Figure 2: SEM picture of a neuronal network at beginning of growth after 3 days on chip. After about 3 weeks the culture area of the chip is completely overgrown by cells.

Methods and Experiments

Living cells are directly cultured on the sensorchip (see Fig. 1&2). Neuronal cells from dissociated tissue cultures were prepared according to the basic method established by Ransom et al. [7]. The network developed spontaneous electrical activity after about one week and stabilized after three weeks on chip. The culture area on the chip is less than 20 mm^2 and has a chamber volume of 10 µl in the flow injection system.

On our present CMOS sensorchip we integrated a multi electrode array (MEA) with 58 palladium electrodes, temperature sensor, ion sensitive field effect transistors (ISFETs) and two types of oxygen sensors. Therewith we can non-invasively measure metabolic parameters like acidification and respiration rate of the cells. As electrical signals we can additionally observe the action potential signals of neuronal networks cultured on our CMOS biosensorchip (see Fig. 3).

Results and Discussion

The electrical activity of neuronal networks from murine fetal spinal cord and frontal cortex tissues was measured with the palladium electrodes (see Fig 3). The results of the electrical activity measurements are comparable with the established multi electrode arrays (MEA) on glass chips [8].

Measurements of acidification rate and oxygen consumption have been made with the neurochip and with the sensor system from the Bionas GmbH. With the Bionas sensor chip the adhesion could be measured as an additional parameter.

Conclusion and Outlook

The more parameters we can study in parallel the better we will understand for example cellular reactions in cellular biosensors or drug screening applications. We realized a CMS with a silicon based biosensorchip for the measurement of the electrical as well as metabolic activity of neuronal networks which can be used as an detector system for bio-toxic agents.

At present we test a CMOS sensorchip prototype with integrated electronics (filters, preamplifiers, multiplexer, I^{2}C-bus, …). With the integrated electronics more sensors

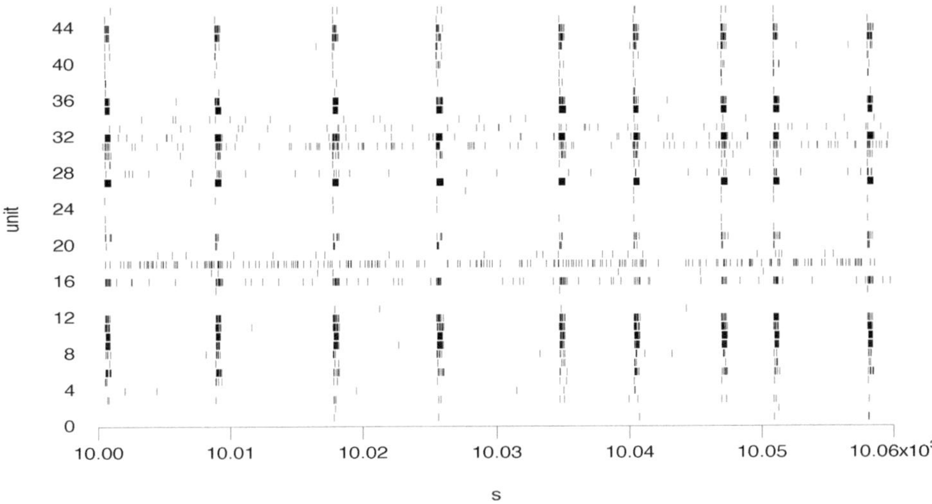

Figure 3: Cutout (60 sec) of 46 spike trains from different electrodes of a spontaneous active frontal cortex network after 4 weeks in culture on the chip. Each bar in a spike train line represents one action potential. The pattern is specific to the used cell type as well as to the reaction to added neuronal active substances.

can be integrated using the same chip carrier and additional features like PC controlled switching between stimulation and recording can be realized. Also the complete system setup can therewith be significantly miniaturized.

To manage the high amount of data we also develop algorithms for automated spike detection and unit separation. For the interpretation of the complex neuronal data sets pattern recognition methods are used.

Acknowledgement

The research effort is sponsored by the European Regional Development Fund (ERDF), the state Mecklenburg-Vorpommern, the Micronas GmbH and the Bionas GmbH.

References

[1] Baumann, W., et al., *Sensors and Actuators B*, B 55 (1999), 77-89
[2] Ehret, R., et al., *Fresenius J Anal Chem*, 369 (2001), 30-35
[3] Baumann, W., et al., Proceedings µ-TAS, Malmö, Sept. 2004, Vol. II, 554-556
[4] Krause, G., et al., *Biosensors & Bioelectronics*, (2006) 1272-1282
[5] Henning, T., et al., *Anti-Cancer Drugs*, 12 (2001), 21-32
[6] Lehmann, M., et al., *Experimental Cell Research*, Vol 305/2 (2005), 374-382.
[7] Ransom B.R., et al., *J. Neurophysiol.*, 40 (1977) 1132- 1150
[8] Gross G.W., et al., *Biosensors & Bioelectronics*, 10 (1995), 553-567

L.M. Lechuga et al. (Eds.)
IOS Press, 2008

Cell-based analyzing system for continuous determination of cell physiology

Elke THEDINGA*, Sabine DRECHSLER*, Axel KOB*, Marcus WEGO*, Miriam NICKEL*, Werner BAUMANN#, Ingo FREUND**, Mirko LEHMANN**, Ralf EHRET*
*Bionas GmbH, Friedrich-Barnewitz-Str. 3, D-18119 Rostock Germany
#Biophysics Institute, University Rostock, Gertrudenstr. 11A, D-18057 Rostock, Germany;
**Micronas GmbH, Hans-Bunte-Str. 19, D-79108 Freiburg, Germany.

Abstract. To characterize modes of action of substances as well as their cytotoxic effects Bionas GmbH has developed a new screening system to allow the continuous recording of how an active substance can act (Bionas® 2500 *analyzing system*). In the pharmaceutical and chemical industry as well as in environmental science it is important to acquire as much information as possible about the metabolic effects of an active substance or there cytotoxicity. With the Bionas® 2500 *analyzing system* metabolically relevant data including oxygen consumption, acidification rate and the adhesion (cell impedance) of cells can be measured in parallel, on-line and label-free. Using e.g. ion-sensitive field effect-transistors (ISFET) and electrode structures it is possible to observe metabolic parameters non-invasively and continuously over longer periods of time. The system has already been established for several cell models, cell lines as well as primary cells. The strength of our system can be found in the continuous data collection during the whole application period. Dynamic, reversible and / or regenerative processes can be observed in one and the same culture. Adaptation effects through repeated addition of compounds can also be observed. A long application period (hours to days) allows more realistic compound concentrations as it is possible in short-term experiments. This advantages result in a higher information content compared to end-point-methods. It also offers the advantage that regenerative effects can be observed during the same test run.

Keywords: label-free, non-invasive, cell metabolism, in vitro, toxicology, drinking water evaluation, on-line monitoring, extracellular acidification, oxygen consumption, cell impedance, adhesion, confluence, pharmacodynamic, regeneration, recovery.

Introduction

High quality of drinking water is important to protect public health. Comprehensive identification of chemical contaminants in e.g. water supplies can be a lengthy process, but rapid analytical methods suitable for field use are limited. A complementary approach is to directly measure toxicity instead of individual chemical constituents.

Furthermore, the goal of pharmaceutical industry is to bring forward novel drugs that provide true health benefit to patients via optimized efficacy and minimal adverse effects. Pharmacodynamics is the study of the biochemical and physiological effects of drugs and the mechanisms of drug action and the relationship between drug concentration and effect. Studies on the mode of action and/or effects of a substance related or not related to its desired (therapeutic) target are important topics of the risk assessment.

Frequently, so-called end-point in vitro tests are used, which require many individual tests before a time curve of the compound effect can be analyzed. Advanced in vitro methods may help to clarify cytotoxic or metabolic properties of substances and accelerate the testing process and reduce the costs.

Therefore, a promising approach is the functional online analysis of living cells in physiologically controlled environments for extended periods of time. Parallel and on-line acquisition of data, related to different cellular targets, seems to be required for advanced stages of drug screening and for economizing animal tests.

A system capable of measuring changes in the extracellular acidification rate was available from Molecular Devices Inc. [Cytosensor™ Microphysiometer, 1-3].

In this article, we report about the Bionas® 2500 *analyzing system* which is able to measure label-free the extracellular acidification, the oxygen consumption and the cell adhesion (cell impedance) over extended periods of time. Changes in these parameters in response to test compounds reflect effects on cell metabolism [1-8]. As an additional advantage, regeneration and recovery effects can be monitored. Using different cell types from different species it is therefore possible to create in vitro assays which closely reflect the in vivo situation.

Physiological relevance

An overview of cell metabolism and the measurement parameters (Fig. 1) shows the physiological relevance of the readouts.

Energy metabolism

Almost any metabolic process used to generate energy (ATP generation) in the cells is defined as cell respiration or "internal respiration". Cells (eukaryotic) take in carbohydrates for their energy supply (e.g. glucose) which are then broken down in the cytoplasm and mitochondria to carbon dioxide and water. During this process a range of reaction steps are traversed, including many redox reactions, that enable the transport of hydrogen (protons and electrons) to the mitochondria by transferring hydrogen atoms to coenzymes (e.g. NAD^+). Within the mitochondrial respiratory chain hydrogen is then transferred to oxygen, a process associated with the generation of ATP. Ultimately, a complete oxidation of the carbohydrates occurs, a process which can be subdivided in four steps:

- glycolysis,
- oxidative decarboxylation,
- citric acid cycle and
- oxidative phosphorylation and end oxidation in the respiratory chain.

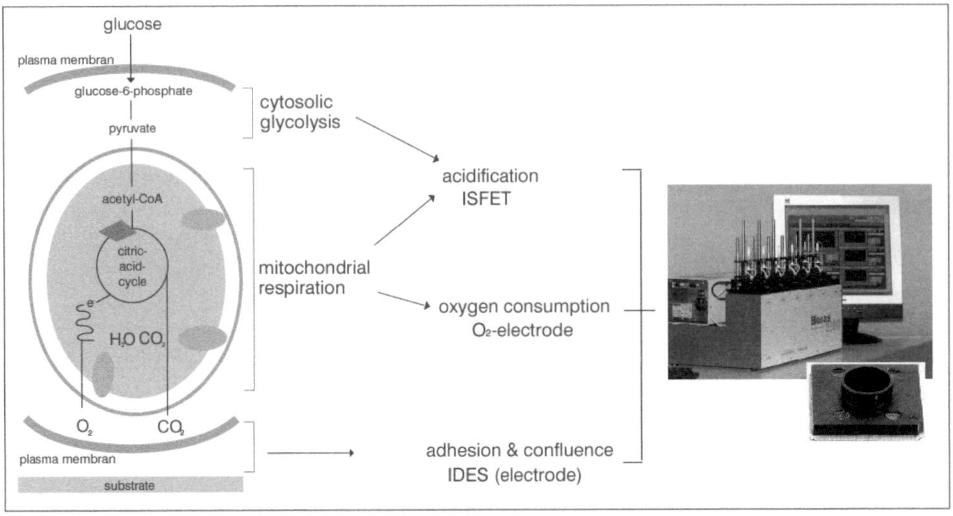

Fig. 1. Cell metabolism and the parameters which can be measured by Bionas®2500 analyzing system.

Glucose metabolism / Acidification activity

Glycolysis occurs in the cytosol of the cells. Glucose is metabolized under aerobic conditions to acetyl-CoA and further to CO_2 (in the citrate cycle). In prokaryotes and in mainly anaerobically active eukaryotic cells or tissues (skeletal muscle) glucose is metabolized anaerobically to lactic acid. So, the breakdown products of glycolysis are lactate and carbon dioxide. In vivo, lactate is either metabolized in the liver and CO_2 expired via the respiratory organs. In vitro (in cell cultures) lactate either passively diffuses through the membrane as lactic acid or is discharged via a carrier. CO_2 is led either passively or actively via anion transport systems through the membrane. It is then hydrated to form carbonic acid after it is liberated into the surrounding solution ($H^+ + HCO_3^-$). Both lactic acid and carbon dioxide are weak acids that contribute to extracellular acidification.

Glycolysis and the subsequent processing in the citrate acid cycle therefore make the greatest contribution to the acidification of the surrounding medium. Furthermore glycolysis in cells in vitro (cell cultures) is also more active than it is in vivo.

Oxidative phosphorylation / Oxygen consumption

In addition to the breakdown products of glycolysis and citrate acid cycle, the consumption of oxygen is also a characteristic feature of cellular metabolism. The main objective of cell respiration is to generate energy in the form of ATP. The production of ATP as a universal "currency" for a large number of metabolic processes mainly occurs in the mitochondria by oxidative phosphorylation. Within the respiratory chain (oxidative phosphorylation) ATP is generated as a chemically storable energy form by the step-by-step transfer of protons (H^+) and electrons (e^-) to elementary oxygen. The reduction equivalents (protons and electrons) supplied by NADH + H^+, $FMNH_2$ and $FADH_2$ originate from the citric acid cycle, the breakdown of fatty acids and from the glycolysis. For each molecule of ATP synthesized half an O_2 molecule is used. The elementary oxygen is taken up by the cells (in vitro) from the surrounding medium and gains access to the mitochondria via diffusion. Therefore, a perceptible oxygen deficit arises in the surrounding medium.

Energy consuming activities

The activation of energy consuming reactions leads to a raised consumption of oxygen and reduction equivalents. Cellular reactions such as signal transduction and stimulation of membrane bound receptors (G protein bound, receptors bound to tyrosine kinase, ion channels) require energy in terms of ATP. The steps of the reaction cascades are directly or indirectly dependent on ATP.

The opening of ion channels for example leads to energy consuming reactions. Ions must be actively and continuously transported through the cell membrane in order to regulate intracellular ion concentrations and maintain hydrostatic equilibrium. This active transportation (Na^+/K^+ ATPases) requires ATP. In the normal case approx. 10^8 protons / second / per cell are transported through the membrane. After stimulation of a receptor this amount can increase by up to 100 % (depending on cell types, receptors and associated activations). In order to achieve such an increase in active proton exchange the cell must synthesize 10^7-10^8 ATP molecules so that it can operate the pumps [1-3]. The consumption of ATP in such large amounts requires an increased delivery of reduction equivalents from glycolysis, citric acid cycle and oxidative decarboxylation, as well as a raised need for oxygen. The ATP consuming pump activities are therefore de facto associated with an increased activation of metabolism. This in turn leads to an increased release of acidic breakdown products and an increasing oxygen deficit in the medium.

In nerve cells the pumps also assume a special role in the transmission of nervous impulses. It is estimated that approx. 70 % of the entire cytoplasmic ATP in nerve cells is exhausted by this process. Ion pumps therefore appear to represent the main component of all energy consuming reactions. The production of the secondary messenger cAMP and phosphorylations are comparatively "cheap" processes that lead to a smaller acidification and a lower oxygen consumption.

In vivo, sugar, amino acids and fatty acids serve above all as carbon suppliers. In vitro (in cell cultures) D-glucose, pyruvate and glutamine are employed as carbon sources.

Cell adhesion and confluence

A further important parameter which describes general properties of the cell is the cell adhesion [10-13]. Adhesion of cells is accomplished by specific adhesion molecules that occur on almost all mammalian cells. Adhesion molecules are proteins that according to receptor-ligand principles establish targeted contacts between cells and in this way provide intercellular communication. The cell-cell contact induces a variety of reactions including the intracellular activation of second messengers and associated events (restructuring and control of the cytoskeleton, formation of surface protein clusters in the cell-cell contact area etc.). Many of the adhesion molecules are only temporarily present on the cell surface. Most are stored in the cell in vesicles and are only released upon external signalling by exocytosis into the surroundings of the cells. Therefore, a temporary increase in the concentration of the adhesion molecules on the cell surface is achieved within best time. This results in a strengthening of cell-cell contacts. After the signal abates together with its associated effects the number of adhesion molecules on the cell surface decreases again.

Adhesion between cells and to substrates represents an important parameter of the cell morphology which is mediated largely via the cytoskeleton. The measurement of adhesion therefore represents a very important parameter for determining cellular behavior.

Detection System

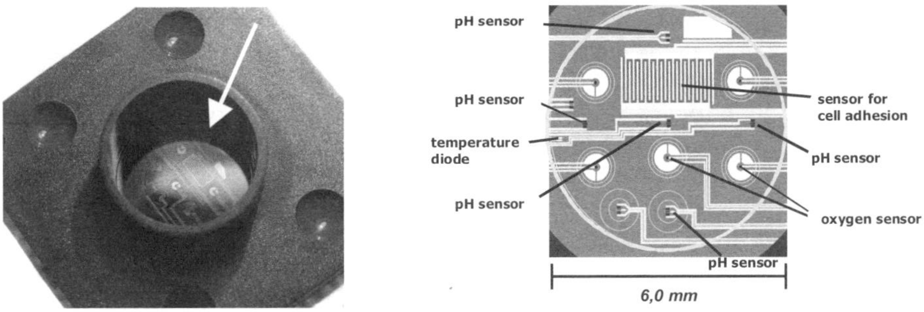

Fig.: 2 a+b. Picture and layout of sensor chip

Online monitoring was performed with the Bionas® 2500 *analyzing system* (Fig. 1). The silicon sensor chip technology allows a continuous recording of cell physiological parameters. By use of the Bionas® *metabolic chip* SC1000 (Fig. 2a) acidification rate, oxygen consumption and adhesion (cell impedance) of cells can be determined from each chip in parallel. Each system can accept 6 chips (biomodules). The layout of the sensor chip (Fig. 2b) shows the cell culture and sensor area. Five sensors for pH, two for oxygen, one for cell impedance measurement and a temperature diode are located in the area marked in yellow.

Changes in the pH are measured on the sensor chip by ion sensitive field effect transistors (ISFETs) and oxygen consumption is measured extracellular by a modified Clark type electrode on the sensor surface. The cell adhesion is detected by an IDES-Sensor (interdigitated electrode structures). For this an Impedance measurement is used to detect the presence of electrically insulating (cell) membranes near the measurement electrodes [4-5]. The results (adhesion/ impedance) are influenced by:

- distance from cell membrane to electrodes
- electrode surface covered by cell membranes
- electrical insulating property of cell membranes,
- cell-cell contacts,
- cell shape (morphology),
- conductivity of the solution (hardly changes during cell culture).

Therefore, impedance measurement gives a measure of the morphology and membrane integrity of cells.

A cavity of sensor chip which works directly as a cell culture well is formed by encapsulation with a special epoxy (Fig. 2.a). The cells are seeded directly on the chip surface to assure highly specific signal detection. The measurement is non-invasive and label-free.

A reaction chamber is formed by insertion of sensor chip in the biomodule of system (Fig 3). The height of the reaction chamber above the sensor surface is 200 µm. Supply of fresh medium is achieved by fluidic perfusion of the analyzing system, which is connected to an autosampler and controlled by a PC. Modified, slightly buffered medium (RM, running medium) is fed over the cells at a pump rate of 56µl/min. Test compounds dissolved in medium are also infused by the automated fluid handling system. The breakdown products (lactate, CO_2) released into the medium by the cells as well as the oxygen consumption of the cells result in a change in pH and oxygen content in the medium.

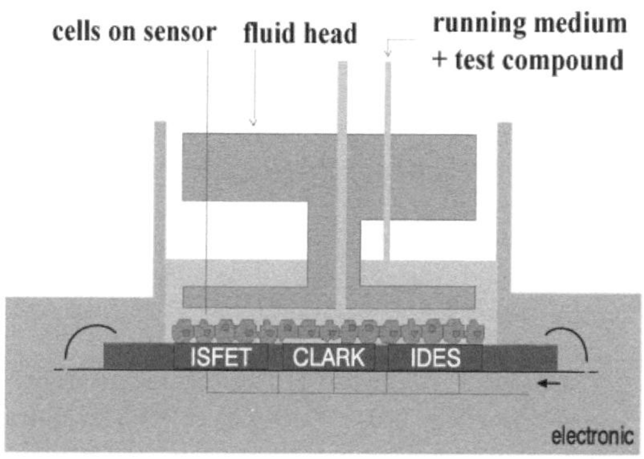

Fig. 3. Diagram of reaction chamber in the biomoduls of the Bionas® 2500 *analyzing system*. Each chamber is connected with pump and control unit

The media flow is stopped periodically. This means the changes of pH and oxygen consumption are measured in the stop phase of the pump cycle [4-8] and in the following pump phase the "used" medium is substituted with fresh medium of a predefined pH and oxygen content (Fig. 4). This pump cycle is carried out during the whole experiment. Typically the stop and pump phases last 3-4 minutes each.. The acidification and respiration rates are calculated by the slope of changes in the stop phase caused by the cell reaction and are standardized to 100% of the basic signal just before substance addition. Impedance measurement is carried out continuously and does not depend on the pump cycle. The output signal (impedance) was also related to the basal level and is given in %.

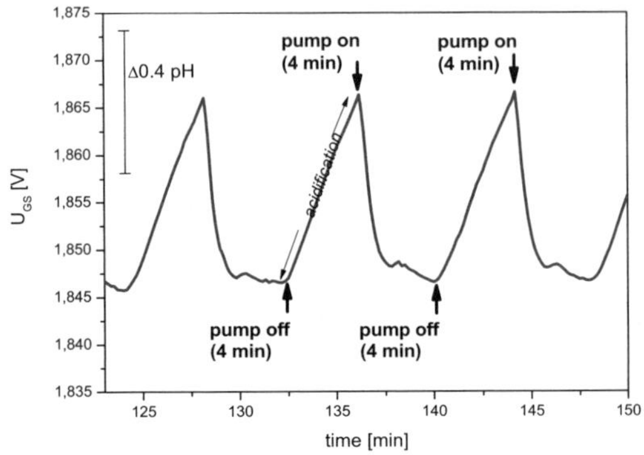

Fig. 4. Measurement mode: Stop and Go phases of fluidic perfusion system (example: acidification measurement)

Materials and methods

Chemicals, media and assay kits

All chemicals were purchased from Sigma-Aldrich (Munich, Germany). Cell culture media were obtained from Cambrex (Verviers, Belgium). RPMI-1640 and MEM Media without sodium carbonate buffer for preparation of running medium (RM) and Poly-L-Lysin were obtained from Biochrom (Berlin, Germany). DMEM without sodium carbonate buffer was obtained from CCPro (Neustadt, Germany). For measurement the pH of media was adjusted to 7.4 and the osmolality to 290 mOsmol/kg.

ATP assay kit (ATPlite™-M Luminescence Assay) was purchased from Perkin Elmer (Rodgau-Jügesheim, Germany) and the Lactate Assay Kit from Biocat (Heidelberg, Germany).

Cells

Cell lines were purchased from Cell Line Service (Eppelheim, Germany).

Seeding of cells on the chip surface

The sensor chips were also disinfected with 70% ethanol, washed with PBS and conditioned with standard medium (RPMI-1640, MEM or DMEM; without FCS containing 10.000 U penicillin and 10 mg streptomycin per ml) over night in a CO2 incubator. Chips for HepG2 cell cultivation were pre-coated with Poly-L-Lysin. The cell lines employed (CHO transfected with actetylcholin receptor, V79, HT-29, HepG2 and TE671) were cultured at a density of approx. 1 - 2 * 10^5 cells / chip (450 µl) in culture medium (RPMI-1640 medium for CHO, V79, HT-29, MEM for HepG2 and DMEM for TE671) supplemented with 10 % FCS, 10.000 U penicillin and 10 mg streptomycin per ml within a CO_2 incubator at 37°C and 5 % CO_2. The cells were detached and resuspended in medium. The cell suspension was directly pipetted on the chip surface and the cells were cultured for one (V79, HT-29, CHO; TE671) or two (HepG2) days in cell culture medium within a CO_2 incubator at 37°C and 5 % CO_2. The cell concentration per cell line was adapted in such a way that the cells reach approx. 80 % confluence on the sensor chips after 24h or 48 h. For comparison of our online analysis with conventional end point assays, the HepG2 cells and the HT-29 cells were also cultured in conventional two dimensional systems. In these cultures release of lactate was determined and also the mitochondrial activity (MTT-Test) and the intracellular ATP content.

Preparation of system and implementation of measurement

Before each experiment the tubing system of the Bionas® 2500 *analyzing system* (Fig. 1) was disinfected with 70 % ethanol and then rinsed with PBS and conditioned with respective running medium (medium without sodium carbonate buffer containing 1mM Hepes, 0.1% FCS, 10.000 U penicillin and 10 mg streptomycin per ml). The pH of the running medium was adjusted to 7.4 and the osmolality to 290-340 mOsmol/kg.

After preparation of the Bionas® 2500 *analyzing system* cell-chips were taken from the incubator and brought to the biomoduls of the system before start.

As a first step the base lines for acidification, respiration and cell adhesion (impedance) were determined. After a stabilization phase of about 3-4h the addition of test substances followed and their effects on the cells were measured over an incubation time of 2h (Fig. 5-9), 3h (Fig. 10-13; Fig. 21-26) and 24h (Fig. 16-19). By removing the active substance a regeneration phase only with running medium was attached. At the end of the experiment the cells were killed by addition of 0.2 % Triton X-100 to the running medium to get a basic signal without living cells on the sensor surfaces (negative control).

For the graphic presentation 5 measured values just before application of the test substances were averaged and standardized to 100 %, while the values from the cells killed with Triton X-100 addition were set to 0 %.

MTT-test

Evaluation of mitochondrial activity was performed by MTT assay [14]. HepG2 cells were seeded in cell culture medium with 10% FCS at a density of 1 x 10^5 cells/well in 96-well plates and incubated for 24 h. Before addition of compounds the medium was exchanged against fresh medium with 1% DMSO and 0.1 % FCS. After 2h medium was exchanged against fresh medium (0,1 % FCS) with compound (Diclofenac) or without compound (control) and final DMSO concentration of 1%. Incubation for 3h was performed in a CO_2 incubator at 37°C. After incubations 50 µl MTT (5mg/ml) in PBS was added for 1-1,5h. MTT-containing medium was removed and 200 µl of DMSO was added to dissolve the formazan crystal formed by live cells. Absorbance was measured at 570 nm.

ATP assay

Cells were seeded at a density of 1 x 10^5 cells/well in 96-well plates in cell culture medium. After incubation over night the medium was exchanged against fresh cell culture medium with 1% DMSO and 0.1 % FCS for at least 2h. The prepared compound solutions (Diclofenac) or control solution were added into cell cultures in the presence of 0.1 % FBS and final DMSO concentration as motioned above. ATPlite™-M Luminescence Assay was performed according to there instructions of the assay kits.

Lactate assay

HT-29 cells were seeded in cell culture medium with 10% FCS at a density of 1 x 10^5 cells/well in 96-well plates and incubated for 24 h. Before addition of compounds the medium was exchanged against fresh medium with 0,005% DMSO and 0.1 % FCS. After 2h medium was exchanged against fresh medium (0,1 % FCS) with compound (Cycloheximide) or without compound (control) and final DMSO concentration of 0,005%. Incubation was performed in a CO_2 incubator at 37°C for 24h. Additionally, after incubation the medium with or without compounds was exchanged against fresh medium without compound and incubated for 2,5h. For Lactate detection 10µl medium of each well was used. Lactate-Assay-Kit was performed according to the instructions from Biocat.

Results

Biological mechanisms of action

For demonstrative purposes various substances with known mechanisms of action were selected in order to illustrate the sensitivity of the system (Fig. 5-9). The effects of the substances on the cells could be clearly recorded:

a) NaF on CHO cells (Fig. 5) inhibits glycolysis, as evidenced by a reduction in the acidification rate. The adhesion (cell impedance) is slightly increased. In addition, regeneration can be seen.

b) NaN3 (Fig. 6) blocks cytochrome-C oxidase and thereby inhibits the respiratory chain of CHO cells. The respiratory rate is clearly reduced in the measurement, while the acidification rate is not influenced and the adhesion (cell impedance) is only slightly influenced. Again, regeneration can be seen.

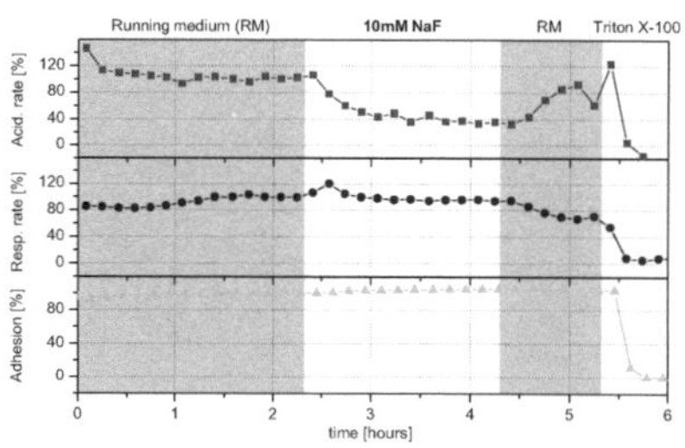

Fig. 5. Measurement with 10 mM NaF (CHOcells).

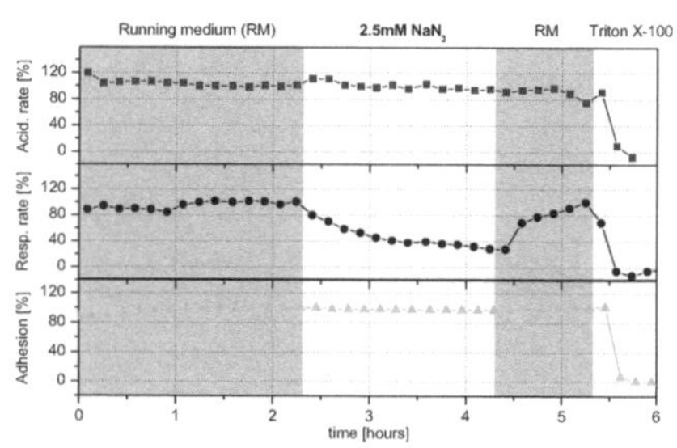

Fig. 6. Measurement with 10 mM NaN3 (CHO cells).

c) KCN (Fig. 7) blocks the cytochrome-C-oxidase and in this way inhibits the respiratory chain. The respiratory rate is rapidly reduced to 0% in CHO-cells while the acidification rate is slightly increased and the adhesion (cell impedance) is only slightly influenced. In this case, no regeneration can be seen, and the reaction is therefore not reversible. The values for the acidification rate and the adhesion hardly show any changes during the regeneration phase.

d) Cytochalasin B on V79 cells (Fig. 8) inhibits both glucose transports through the membrane as well as cellular proliferation by blocking actin polymerization. The acidification rate is reduced slightly, while the respiration is increased slightly. Adhesion (cell impedance) is, however, clearly reduced.

e) The mode of action for acetaminophen (Fig. 9) is the inhibition of cyclooxigenase (COX) but not precisely clarified. It is remarkable, however, that despite the high concentration of 37 mM (IC_{50} approx. 19 mM, [15]) a clear regeneration can be seen in CHO cells.

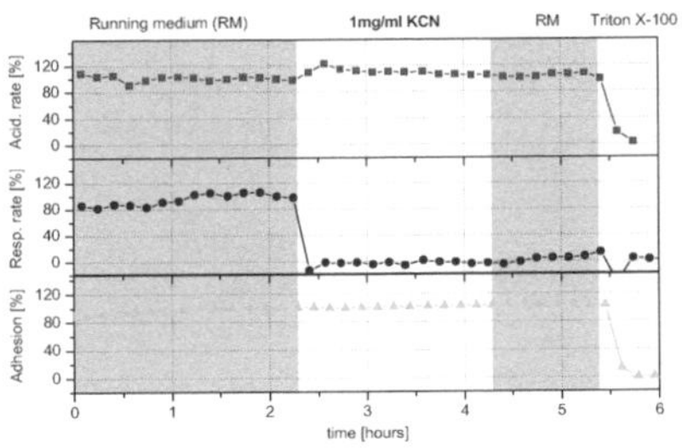

Fig. 7. Measurement with 1mg/ml KCN (CHO cells).

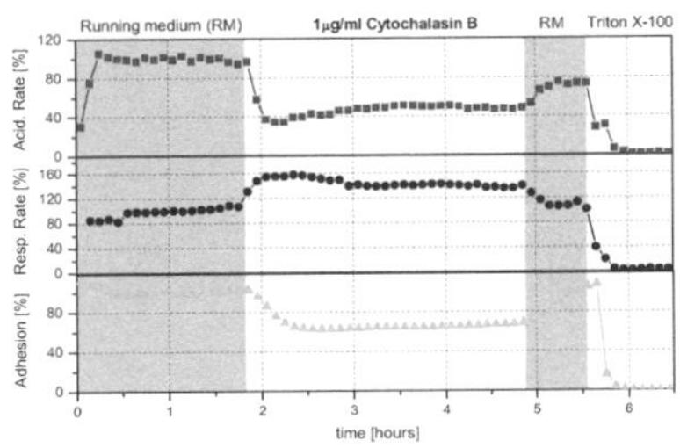

Fig. 8. Measurement with 1µg/ml cytochalasin B (V79 cells).

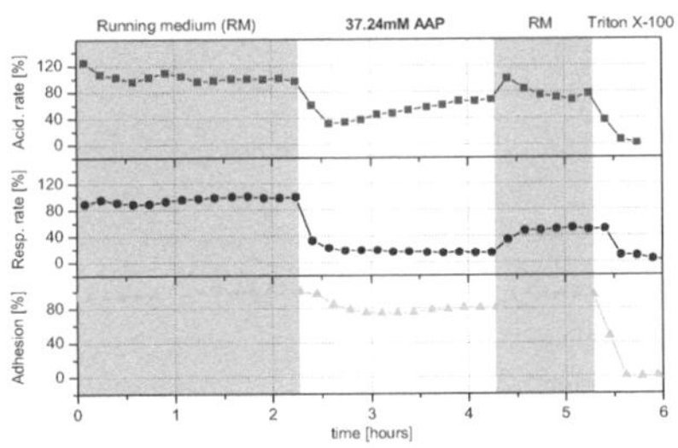

Fig. 9. Measurement with 37mM acetaminophen (CHO cells).

Dynamic effects of compounds

To demonstrate the advantage of the time resolved online monitoring we used Diclofenac as test compound. The human liver cell line HepG2 was exposured to 0.1mM, 1mM, 5mM and 10mM Diclofenac over 3h. Diclofenac is one of the world's most popular nonsteroidal anti-inflammatory drugs (NSAID) which inhibits Cyclooxigenase (COX) activity. The measurement in the Bionas® 2500 *analyzing system* shows a concentration depended increase of the acidification rate at the beginning of the exposure time with Diclofenac (Fig. 10). After the first peak a rapid decrease of acidification activity is visible at the 5mM and 10mM Diclofenac concentration whereas 1mM and 0.1mM resulted only in a slight and slow decrease.

Interestingly, after removing Diclofenac out from the medium (second grey field: RM) only the lowest Diclofenac concentrations (0.1mM, 1mM) show regeneration of cell metabolism. The highest concentrations of Diclofenac caused a remaining damage. No regeneration could be observed.

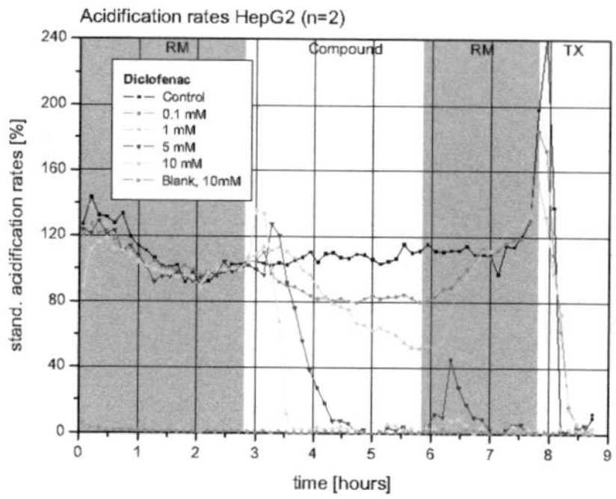

Fig. 10. Standardized acidification rate of HepG2 cells influenced by Diclofenac (n=2).

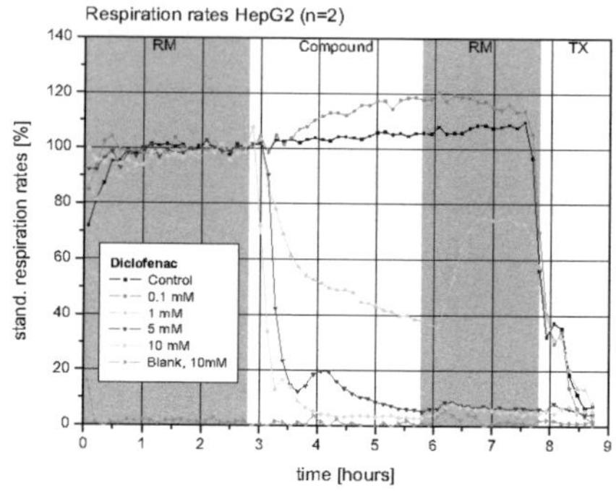

Fig. 11. Standardized respiration rate of HepG2 cells exposured to Diclofenac (n=2).

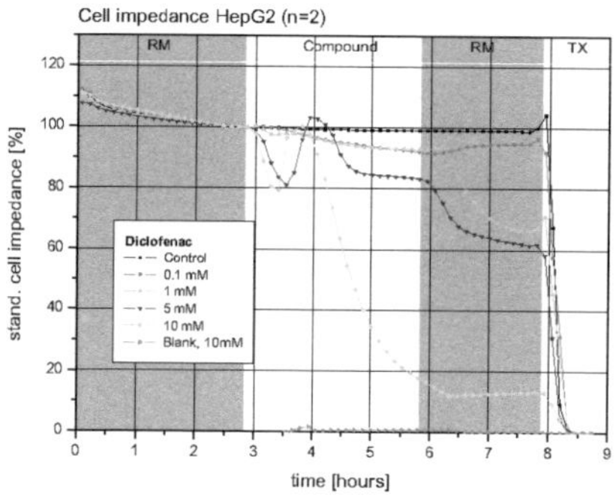

Fig. 12. Standardized cell adhesion of HepG2 cells exposed to Diclofenac (n=2).

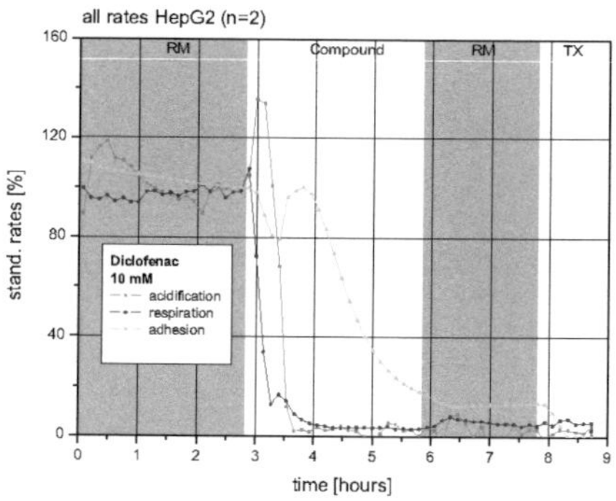

Fig. 13. Standardized acidification rate, respiration rate and cell adhesion of HepG2 cells exposed to 10mM Diclofenac (n=2).

The 0.1mM Diclofenac concentration resulted in a slight increase of the respiration rate compared to the control (Fig. 11). By 1mM Diclofenac respiration activity was reduced to 40% by and almost completely by 5mM and 10mM. Regeneration effects can only observed for 1mM Diclofenac concentration (second grey field of Fig. 11)

The cell adhesion (Fig. 12) shows very interesting effects at high Diclofenac concentrations (5mM, 10mM). First cell adhesion was reduced by exposure to 5mM and 10mM Diclofenac for approx. 30 min. After then, the cell adhesion was enhanced also for only approx. 30 min followed by an anew decrease. Finally, 10mM Diclofenac led to decrease of 82%, 5mM to 18% and 1mM as well as 0.1mM Diclofenac only to approx. 8%. In difference to acidification activity and respiration behavior cell adhesion did not show regeneration effects after removing Diclofenac out from the medium.

By displaying acidification rate, respiration rate and cell adhesion measurement of HepG2 cells in one graph (Fig. 13) dynamic effects of cell behavior influenced by

Diclofenac can be seen. First an initial increase of acidification is simultaneous with a decrease of respiration and a first decrease in adhesion. After approx. 30 min acidification rates decreased whereas the adhesion increased at the same time. The respiration rates show only a slight transient increase. Subsequently a new decrease of cell adhesion can be observed.

One hour after Diclofenac incubation there is only a very low activity of acidification and respiration, whereas cell impedance (adhesion) shows nearly 100%. This means, that there is a confluent layer of intact cells (membranes) on the chip, whereas the metabolic activity is drastically reduced (3% acidification, 7% respiration). After removing the compound from medium acidification rate, respiration rate and cell adhesion did not show regeneration effects.

To correlate this observation we measured cellular ATP concentration (Fig. 14) as well as mitochondrial activity by MTT-test (Fig. 15). Both results were diagrammed in relation to control (100%). According to the respiration rate measured in the Bionas® 2500 *analyzing system* intracellular ATP concentration increased at 0,1mM Diclofenac. 1mM Diclofenac led to 30% decrease of cellular ATP concentration compare to control. Diclofenac concentrations above 1mM showed not measurable ATP content. At these concentrations ATP concentration decreased to zero.

The mitochondrial activity measured by MTT-test (Fig. 14) showed an increase for 0,1mM Diclofenac up to 115% and for 1mM almost 190%. Diclofenac above these concentrations led to a fast decrease of mitochondrial activity.

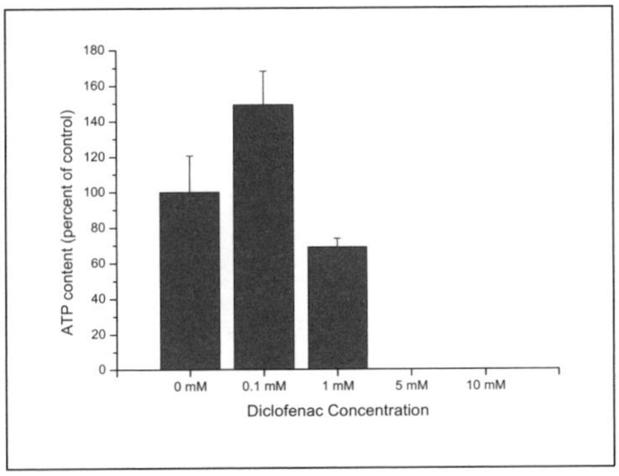

Fig. 14. ATP content (percent of control) after 3h of Diclofenac exposure on HepG2 cells

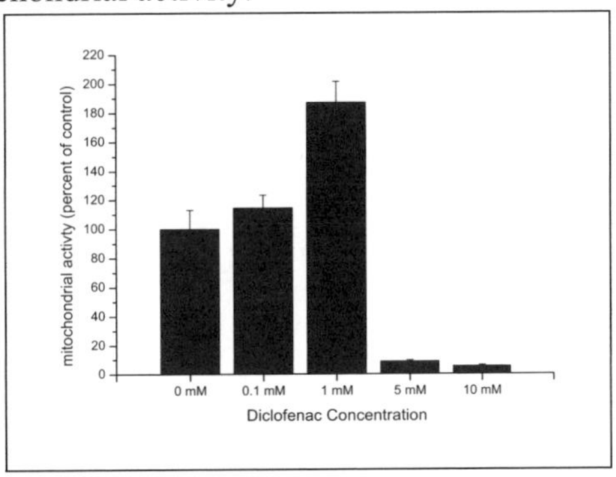

Fig. 15. Mitochondrial activity measured by MTT-test after 3h of Diclofenac exposure on HepG2 cells

For further demonstrative purposes an experiment with Cycloheximide was performed (inhibits eukaryotic protein synthesis by blocking translation). The reduction in protein synthesis becomes recognizable from decrease in acidification and respiration rates (Fig. 16-18). Interestingly, cell adhesion is increased during this phase (Fig. 18). Another interesting effect is the increase of respiration rate of control whereas the acidification rate decreased. This seems to be caused by addition of DMSO, which consist in all medium solution in the same concentration. We have observed this behavior also in some other experiments with HT-29 cells in which we used DMSO in media (Data not shown).

Quite clearly an overshoot of reaction (acidification rate, respiration rate) occurs after removal of blocking effect. Surprisingly the effects in individual parameters do not coincide exactly with each other (Fig. 19). After removing Cycloheximide out from the medium (second grey field of Fig. 17 "Running medium") an initial increase of respiration rate could been observed. Before the respiration rate decreased again the acidification rate had started to increase. The subsequent actions of cell metabolism led to the assumption that the mitochondrial activity was influenced first after removing Cycloheximide out from the medium. The increasing of respiration rate reflects high energy consumption when the cells start to produce proteins again. As a result of the energy consumption the glycolysis was also enhanced with a delay.

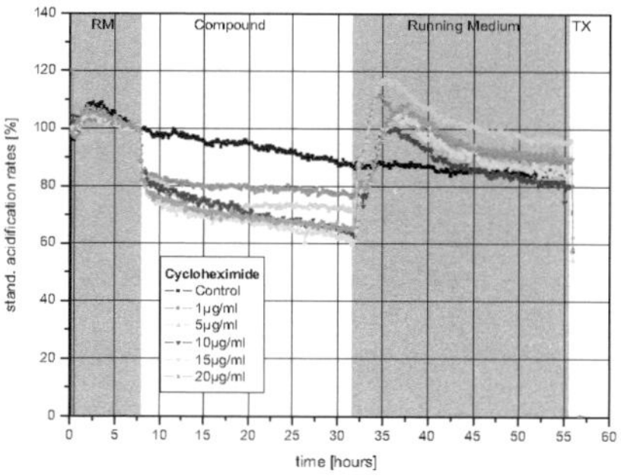

Fig. 16. Standardized acidification rate of HT-29 cells for various concentrations of Cycloheximide (n=3).

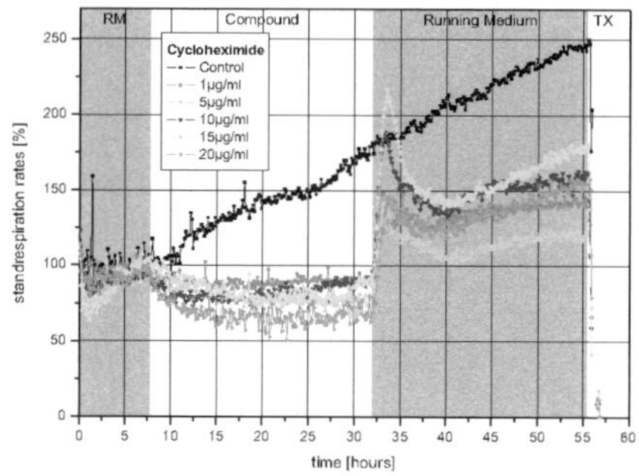

Fig. 17. Standardized respiration rates of HT-29 cells for various concentrations of Cycloheximide (n=3)).

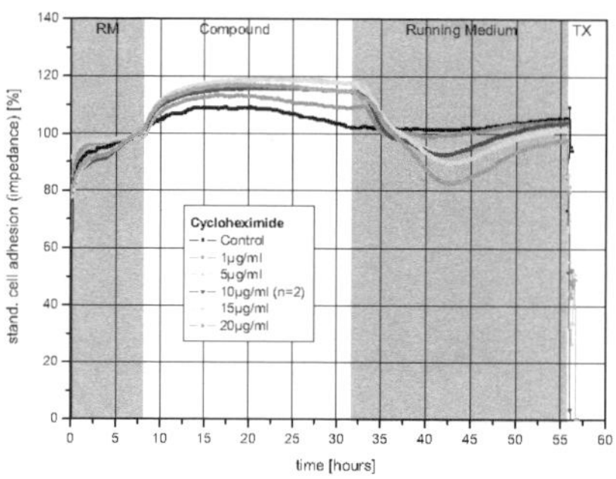

Fig. 18. Standardized cell adhesion (cell impedance) of HT-29 cells for various concentrations of Cycloheximide (n=3, except 10µg/ml (n=2)).

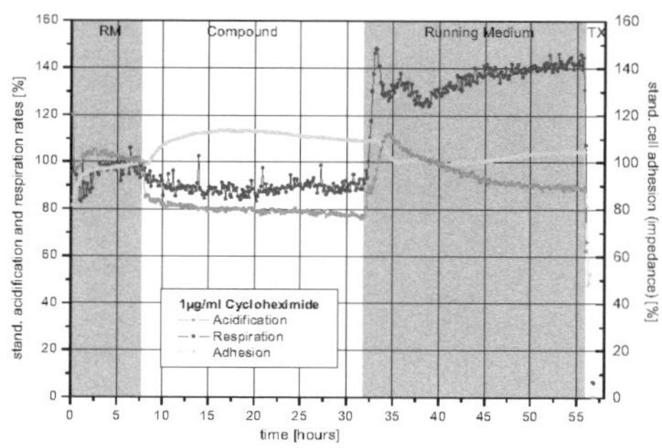

Fig. 19. Acidification, respiration rates and adhesion (cell impedance) after addition of Cycloheximide (1µg/ml) (HT-29 cells).

According to the transient increase of acidification rate we have also detected an increase of lactate release 2,5h after removing of Cycloheximide from the medium (Fig. 20) for 1µg/ml and 5µg/ml Cycloheximide, but not above these concentrations.

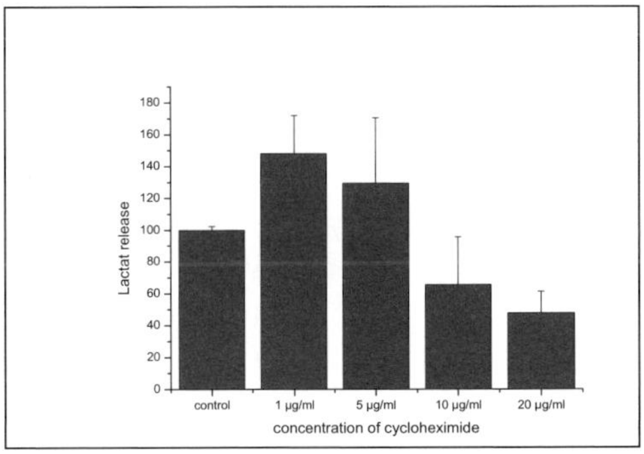

Fig. 20. Lactate release measured by Lactate assay kit after 2,5 h of regeneration (without compound) in HT-29 cells

By addition of Carbachol or EGF to TE-671 cells different receptors were stimulated. In experiments with Carbachol after a first measurement phase with running medium only Carbachol in different concentrations was incubated two times for 1h with a break of 2h between the incubation phases (Fig. 21-23). Receptor activation by Carbachol was visible as an increasing of the acidification rate (Fig. 21). The respiration rate (Fig. 22) was not affected by Carbachol. The cell impedance (Fig. 23) showed a fast increase of rates directly after addition of Carbachol followed by a slow decrease after the maximum value. After withdrawal of compound the cell impedance showed a fast decrease.

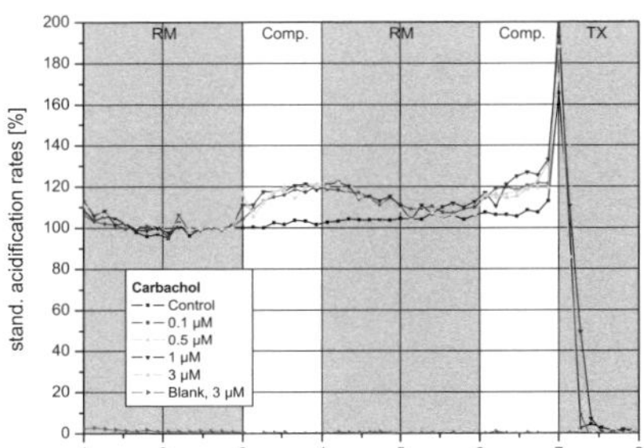

Fig. 21. Standardized acidification rate of TE-671 cells for various concentrations of Carbachol.

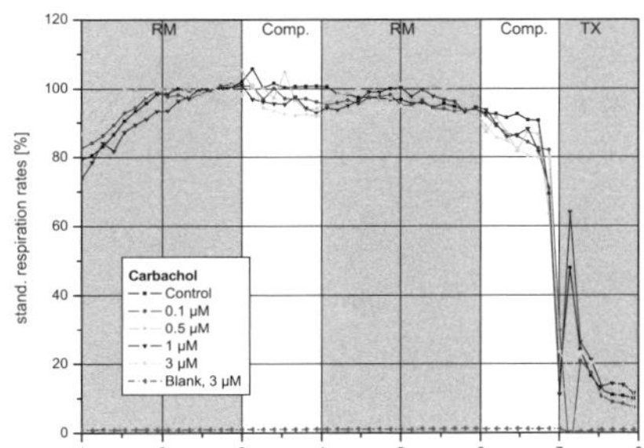

Fig. 22. Standardized respiration rate of TE-671 cells for various concentrations of Carbachol.

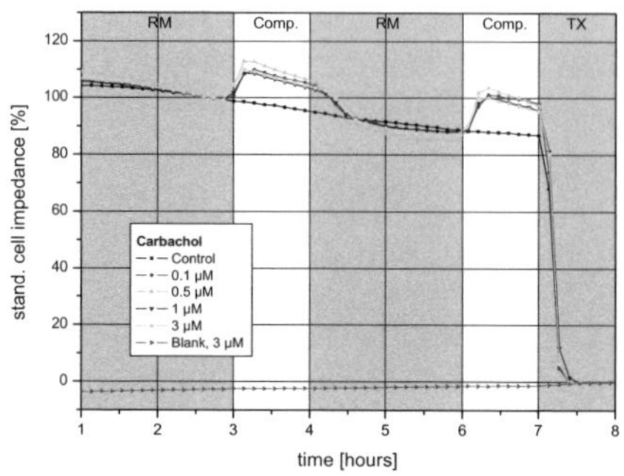

Fig. 23. Standardized adhesion (cell impedance) of TE-671 cells for various concentrations of Carbachol.

EGF incubation on TE-671 cell showed similar effects. In this experiment a 1h incubation phase with two different concentrations was preformed. The 1ng/ml EGF concentration showed the same increase of acidification rate and cell impedance like 3µM Carbachol. 0.1ng/ml EGF resulted in a reduced increase of acidification rates and cell impedance.

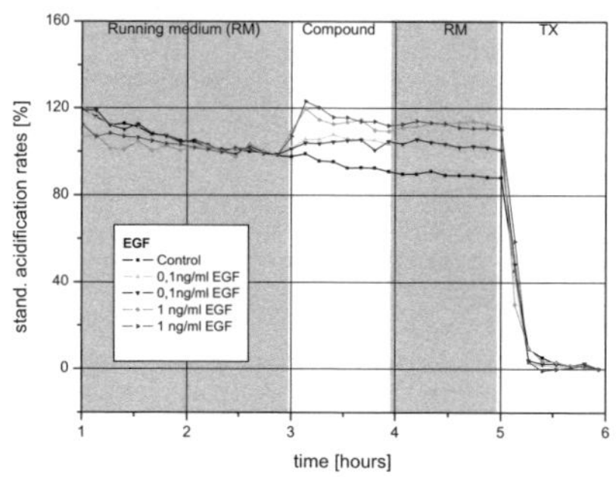

Fig. 24. Standardized acidification rate of TE-671 cells for various concentrations of EGF.

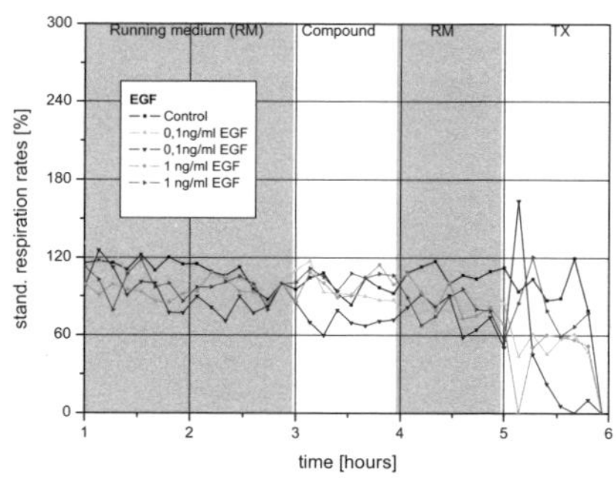

Fig. 25. Standardized respiration rate of TE-671 cells for various concentrations of EGF.

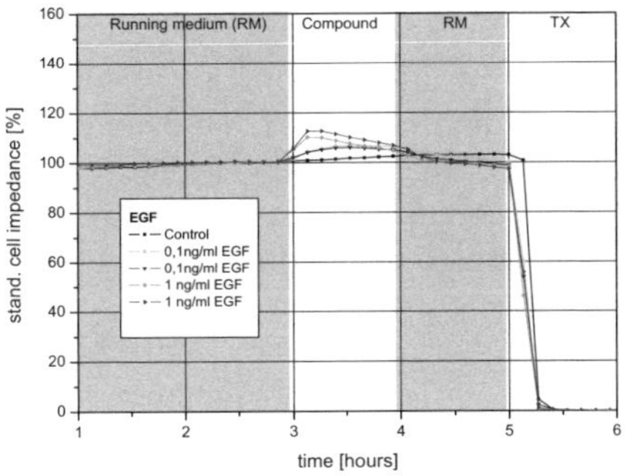

Fig. 26: Standardized adhesion (cell impedance) of TE-671 cells for various concentrations of EGF.

Discussion

By inhibition of single metabolic pathways, we can show the specificity of the signal detection. For example an inhibition of glycolysis, shown in experiments with sodium fluoride (Fig. 5) resulted in a decrease of acidification rates. The inhibition of glycolysis led to a reduction of excretion of acidic breakdown products from cells. Therefore, it resulted in a decrease of acidification activity. During the exposure time (2h) no other metabolic pathways seems to be distinctly influenced. The decrease of respiration rate in the regeneration phase can be interpreted as a subsequent damage of the cells. In similarity to the acidification activity measurement, a selective signal could also be observed by the inhibition of oxidative phosphorylation with sodium azide (Fig. 6) as well as with potassium cyanide (Fig. 7). Only the oxygen consumption is affected by addition of these compounds. In both experiments the cells reduced the oxygen uptake. Except the initial small peak in the acidification rate measurement, no influence of the two other measured parameters is visible. We observed regenerations effects after removing sodium azide from the medium, whereas the cells with potassium cyanide retained damaged.

The experiment with cytochalasin B demonstrated an example for the inhibition of glycloysis and adhesion. Cytochalasin B inhibited the glucose transport through the membrane and influenced the cytoskeleton by inhibition of actin polymerization. With the Bionas® 2500 *analyzing system* both effects can be observed. A reduced acidification rate and a decrease of cell impedance resulted by addition of cytochalasin B. Acetaminophen and Diclofenac are examples for a test compounds which influenced all three measurement signals. An interesting perspective is the time-resolved display of effects. Thus, conclusions can be drawn about the pharmacodynamic of test compounds. The measurement of respiration rate after 3 hours of Diclofenac exposure correlates (Fig. 11) with the results of the ATP assay (Fig. 14) whereas the mitochondrial activity is different. Results of the MTT-test (Fig. 15) indicated that there is a higher mitochondrial activity at 1mM Diclofenac after 3h incubation then at 0,1mM. The discrepancy between oxygen consumption and mitochondrial activity is already known. The oxygen uptake assays are more sensitive to the impairment of mitochondrial function than e. g. MTT-test [16].

Compared to end point assays more time-resolved information can be obtained. Further we have chosen Cycloheximide to show an example in which the main effect is present after exposure and subsequent removing of the compound out form the medium. Cycloheximide is prevents protein synthesis in eukaryontic cells by inhibition of RNA translations. Here, the removing of Cycloheximide resulted in a higher activity of cell metabolism as an overshoot reaction.

Evidence of receptor binding or activation by addition of EGF or Carbachol (Fig. 21-23, Fig. 24-26) could be detected as an increase of acidification rates and cell impedance (adhesion). It is not clear which event of activation caused changes in cell impedance. For example EGF binds to the epidermal growth factor receptor (EGFR). It is a transmembrane glycoprotein which exists on the cell surface. Binding of EGFR to its ligands leads to autophosphorylation of receptor tyrosine kinase and subsequent activation of signal transduction pathways that are involved in regulating cellular proliferation, differentiation, and survival [17]. Upon activation EGFR undergoes a transition from an inactive monomeric form to an active homo (or hetero-) dimer. EGFR dimerization leads to autophophorylation and stimulation of several signalling pathways. Changing of cell impedance (adhesion) may be a result of dimerization by changing of membrane fluidic or by downstream signalling proteins which can modulate e.g. cell migration, proliferation and adhesion. An increase of acidification is an evidence for an activation of glycolysis i.e. for increasing of acid break down product release.

Carbachol is a choline ester which is known to stimulate the muscarinic receptor and nicotinic receptors. The muscarinic receptor is a membrane bound acetycholine receptor which uses G proteins as their signalling mechanism. Alteration of adhesion could be caused by second messengers whereas the increase of the acidification rate is an indicator for higher glycoloysis activity.

Conclusions

The Bionas® 2500 *analyzing system* allows a label-free and multiparametric on-line monitoring of cellular processes. The measurements can be carried out continuously (up to several days) at a high temporal resolution (minutes). Therefore, it is possible to determine the velocity of compound effects. Compared to conventional end point assays it is much easier to get kinetic parameters. In one and the same experiment transient effects can be detected which is almost impossible for end point tests. Also the time course e.g. onset and duration of an (adverse) effect can be investigated together with dose response relationships and repeated dose studies. Regeneration and adaptation behavior can be analyzed with one and the same cells. This enables e.g. a proper time and dose setting for additional examinations. Focusing on relevant metabolic events will increase their efficiency.

Furthermore, this method can be used to detect different signal transduction pathways. It therefore allows an interpretation of the mode of action of a substance which can be used for compound or toxicity ranking.

For pharmaceutical research and development the method also allows extending and optimizing target validation studies by bringing the targets back into their physiological context.

Future efforts to toxicity sensors suitable for use in drinking water evaluations are to clearly define the intended application for the toxicity sensors (with input from the user community) and define acceptable sensitivity ranges for sensor response based on the intended use(s).

In summary, the new analyzing system complements established in vitro tools commonly used for screening of substances and provides a deeper insight into the dynamic of cell metabolism.

References

[1] McConnell, H.M., Owicki, J.C., Parce J.W., Miller, D.L., Baxter, G.T., Wada, H.G., Pitchford, S.: The Cytosensor Microphysiometer: Biological Applications of Silicon Technology. Science, Vol. 257 (1992) 1906-1912.

[2] Owicki, J.C., Bousse, L. J., Hefeman, D.G., Kirk, G.L., Olson, J.D., Wada, H.G., Parce, J.W.: The Light-Addressable Potentiometric Sensor : Principles and Biological Applications. Annu. Rev. Biophysics. Biomol. Struct. (1994) 23:87-113.

[3] Hafner, F.: Cytosensor® Microphysiometer: technology and recent applications. Biosensors & Biolelectronics 15 (2000) 149-158.

[4] Alberts, B. Bray, D., Lewis, J., Raff, M., Roberts, K., Watson, J. D.: Molekularbiologie der Zelle, 2. Auflage, VCH-Verlagsgesellschaft, Weinheim (1990)

[5] Baumann, W., Lehmann, M., Bitzenhofer, M., Schwinde, A., Brischwein, M., Ehret, R. und Wolf, B.: Microelectronic sensor system for microphysiological application on living cells. Sensors and Actuators B, B55 (1999) 77-89

[6] Lehmann, M., Baumann, W., Brischwein, M., Ehret, R., Kraus, M., Schwinde, A., Bitzenhofer, M., Freund, I., Wolf, B.: Noninvasive measurement of cell membrane associated proton gradients by ion sensitive field effect transistor arrays for microphysiological and bioelectronical applications. Biosensors & Bioelectronics, 15 (3-4) (2000) 117-124

[7] Thedinga E, Kob A, Holst H, Keuer A, Drechsler S, Niendorf R, Baumann W, Freund I, Lehmann M, Ehret R. Online monitoring of cell metabolism for studying pharmacodynamic effects. Toxicol Appl Pharmacol. 2007 Apr 1; 220(1): 33-44. Epub 2007 Jan 3.

[8] Thedinga E, Ullrich A, Drechsler S, Niendorf R, Kob A, Runge D, Keuer A, Freund I, Lehmann M, Ehret R. n vitro System for the prediction of hepatotoxic effects in primary hepatocytes. ALTEX. 2007; 24(1):22-34.

[9] Mestres P, Morguet A, Schmidt W, Kob A, Thedinga E. A new method to assess drug sensitivity on breast tumor acute slices preparation. Ann NY Acad Sci. 2006 Dec; 1091:460-9.

[10] Lehmann, M., Baumann, W., Brischwein, M., Gahle, H.J., Freund, I., Ehret, R., Drechsler, S., Palzer, H., Kleintges, M., Sieben, U., Wolf, B.: Simultaneous measurement of cellular respiration and acidification with a single CMOS ISFET. Biosensors & Bioelectronics, 16/3 (2001) pp 195-203.

[11] Ehret, R., Baumann, W., Brischwein, M., Lehmann, M., Henning, T., Freund, I., Drechsler, S., Friedrich, U., Hubert, M.-L., Motrescu, E., Kob, A., Palzer, H., Wolf, B.: Multiparametric cellular biosensor chips for screening applications. Fresenius Journal of Analytical Chemistry, 369 (2001) 30-35

[12] Ehret, R., Baumann, W., Brischwein, M., Schwinde, A., Stegbauer, K., Wolf, B.: Monitoring of cellular behavior by impedance measurements on interdigitated electrode structures. Biosensors & Bioelectronics, 12 (1) (1997) 29-41

[13] Ehret, R., Baumann, W., Brischwein, M., Schwinde, A., Wolf, B.: On-line control of cellular adhesion with impedance measurements using interdigitate electrode structures. Medical & Biological Engineering & Computing, 36 (1998) 365-370

[14] Ceriotti L., Ponti J., Broggi F., Kob A., Drechsler S., Thedinga E., Colpo P., Sabbioni E., Ehret R., Rossi F. Real-time assessment of cytotoxicity by impedance measurement on a 96-well plate. Sensors and Actuators B (2006), in press, accepted manuscript.

[15] Mosmann T. Rapid colorimetric assay for cellular growth and survival: application to proliferation and cytotoxicity assays. J Immunol Methods 1983, 65: 55-63

[16] Clemedson, C. et al: MEIC Evaluation of Acute Systemic Toxicity, ATLA 24 (1996) 273-311

[17] Hynes J., Hill R., Papkovsky DB. The use of a fluorescence-based oxygen uptake assay in the analysis of cytotoxicity. Toxicol In Vitro 2006, Aug; 20(5): 785-92.

[18] Herbst RS. Review of epidermal growth factor receptor biology, Int J Radiat Oncol Biol Phys. 2004; 59(2 Suppl):21-6.

Biosensor detection of microorganisms based on registration of their metabolic activity and immunoassay

A. N. RESHETILOV [a], P. V. ILIASOV [a], Yu. V. PLEKHANOVA [a],
V. I.SIGAYEV [b], A. D. TOLCHINSKIY [b], N.R.DYADISHCHEV [b], R.V.BOROVIK [b]
[a]*Skryabin Institute of Biochemistry and Physiology of Microorganisms RAS, Russian Federation, 142290, Moscow region, Pushchino, Pr. Nauki 5*
[b]*Research Centre for Toxicology and Hygienic Regulation of Biopreparations, Ministry of Health of the Russian Federation, Russian Federation, 142253, Moscow region, Serpukhov, Lenina 102A*

Abstract. In the present work, two biosensor models intended for determination of microbial cells have been developed. One of them uses an approach related to recognition of specific peculiarities of target cells' metabolism and is able to detect 10^6 cells/sample. At the same time, it ensures the recognition of the detected microorganism only in the case of pure culture or qualitation of target on the background of microflora with the known metabolic pattern. Another model is based on the immunoassay technique and can be used for detection of target microorganism with the lower limit of detection of 10^2 cells/sample. The models' operation conditions have been optimized and preliminary conclusion concerning their application has been drawn.

Keywords. Detection of microorganisms, biosensor, cells' enzymatic activity, immunoassay

Introduction

Detection and quantification of microorganisms is a relevant problem of medicine, biotechnology, agriculture, and a number of other fields of human activity. Conventionally, this problem is solved using microbiological, biochemical and serological tests. However, most of such analyses require special equipment and highly skilled personnel and are performed at large stationary laboratories. Besides, such tests take quite a lot of time: from several hours to several days. Besides, often it is necessary to perform assays in situ with immediate interpretation of results. In view of the above, biosensor devices have been intensively developed in the recent years to minimize the equipment and to simplify the preparation and performance of assays [1-4].

The overwhelming majority of the currently known variants of biosensor detection of microorganisms are based on different modifications of immunoassay and DNA analysis providing high sensitivity, selectivity, and relative simplicity of detection. Immunoassay based sensors permit microbial detection with a lower limit of about 10^2-10^3 cells/ml during 30 min – 2 h [5-10]. Electrochemical, optical and acoustic sensors are used as transducers. Sample concentration and cultivation increase the sensitivity of detection to ~10 cells/ml and the duration of assay increases several times. The limit of sensitivity is usually associated with assay duration and varies from tens of cells to 10^3–10^4 cells/sample; the time of assay may be up to 15–20 h. A significant positive characteristic of immune detection is high sensitivity and specificity. Besides, immunoassay allows the detection of bacteria in complex samples (blood, plasma, urea, or food) with the minimal sample pretreatment. At the same time, its disadvantage is positive assessment of a sample containing also nonliving microbial forms. Hence, it is necessary to combine highly selective immune sensors with the sensors based on metabolic activity of microorganisms, since they allow separation of living cells from nonliving ones. A substantial restriction of immunoassay may be significant time expenditures; nevertheless, they are justified when the matter concerns assaying a set of samples: in this case, the specific time for a single sample assay significantly decreases. Particular examples of the works in this field are considered below.

Table 1 presents characteristics of immunosensors for detection of pathogenic microorganisms. Most of them are characterized by the lower detection limit of about 10^3–10^5 cells and assay time of 20–60 min.

Table 1. Characteristics of immunosensors for detection of pathogenic microorganisms

Microorganism	Type of transducer and receptor	Detection limit	Assay time	Reference
Nesseria meningitidis	LAPS (antigen/antibodies with HP label)	< 1000 cells	20 min	[11]
Yersinia pestis		<1000 cells	30 min	
Y. pestis	Acoustic, antibodies	10^6 cells	45 min	[12]
Shigella dysenteriae		10^6 cells		
Y. pestis	Optical, antibodies with fluorescent label	5 ng/ml	30 min	[13]
Brucella melitensis	LAPS, antibodies with urease label	6000 cells	–	[14]
Newcastle disease virus	Optical (registration of evanescent waves)	5 ng	15 min	[14, 15]
Salmonella typhimurium	Acoustic, antibodies	10^5 cells/ml	5 h	[16]
S. typhimurium	Optical (registration of luminescence), antibodies	10^5 cells/ml	30 min	[17]

Abbreviations: LAPS, light addressed potentiometric sensor; HP, horseradish peroxidase.

There are also quite a number of publications describing biosensor models for detection of nonpathogenic or conditionally pathogenic microorganisms. The concentration of *Clostridium thermocellum* cells is determined in the work [18]. Polyclonal antibodies to *C. thermocellum* were used in the experiments. *Mycobacterium thermoautotrophicum* was a control culture for testing assay specificity. The minimal detected cell concentration was 10^4 cells/ml; the maximal studied value was 10^7 cells/ml. The developed scheme of analysis seems to have prospects for express testing of the content of microorganisms in suspension as it permits using simple instruments for measurement and provides high sensitivity of detection.

Theoretical aspects of application of a standard sandwich scheme of immunoassay followed by verification of results on experimental models of detection of bacterial cells have been considered in the work [19]. The studies of quantitative regularities of interaction of HRP labeled antibodies with *Bacillus* sp. cells were a basis of research directed at optimization and comparative assessment of various methods of immune-enzyme assay of bacteria.

The method of fluorimetric analysis, identification, and estimation of vital functions of pathogenic microorganisms (including spore forms) and exotoxins has been developed. This method allowed the detection of ~20 cells/cm^2 or ~100 cells/50 ml in a few minutes. The accuracy of recognition was provided by pattern recognition algorithms and receptor-specific capture of cells. The method can be applied for identification of nonpathogenic microorganisms and PCR analysis [20].

The system of air state monitoring was developed on the basis of electrochemical immunosensor connected with an aerosol collector. The possibility of detection of *Escherichia coli*, *Salmonella* sp., and *Mycobacterium* sp. was demonstrated. The assay time was no more than 1 h; the lower detection limit was ~5000 cells/ml [21].

The conductometric immunosensor for detection of microorganisms was created. *E. coli* cells were used as a model target organism. The sensor allowed detection of 10^2 cells/ml in 10 min. The sensor was proposed to be used for detection of pathogenic microflora in food industry and medicine, as well as for providing biological safety [22].

The work [23] describes a plant for assessment of the presence and concentration of *S. typhimurium*. The device includes a 96-hole immunological plate with a built-in gold oxygen electrode in each well, containing immobilized antibodies against the target microorganism. The principle of detection is based on registration of the decrease of dissolved oxygen during the growth of immunosorbed bacteria. The plant receives a signal at initial cell concentrations of about 10^1-10^3 CFU/ml. The absence of response to the cells of *E. coli*, *P. aeruginosa*, *Corynebacterium aquaticum*, and *B. subtilis* was demonstrated.

Electrochemical immunosensor for *E. coli* detection on the basis of capillary bioreactor-separator was described. Its algorithm included immobilization of antibodies on the inner surface of reactor capillaries, formation of a sandwich immune complex with target microbial cells and alkaline phosphatase-labeled antibodies; the enzyme label activity was registered by amperometric electrode with co-immobilized tyrosinase and horseradish peroxidase after the addition of phenylphosphate to the channel. Bacterial detection was possible in the range of 10^2–10^6 cells/ml; assay time was 1.5 h [24].

The method of concurrent detection of *E. coli* and *S. typhimurium* cells on the basis of immunomagnetic capture followed by fluorimetry with time resolution was presented. The Eu and Sm compounds forming a fluorescent complex after the treatment with nitrile triacetate and trioctylphosphine oxide were used as a label of antibodies. The assay included a 4.5-h stage of cultivation. The possibility of detection of 1 CFU/g sample has been demonstrated [25].

DNA biosensors are a relatively novel approach to detection and identification of microorganisms. Genetic analysis with hybridization of nucleic acids is a detection procedure alternative to immunoassay; when coupled with target material replication using polymerase chain reaction, it provides high levels of sensitivity and selectivity. In accordance with this principle, a microgravimetric sensor has been developed and applied for detection of *S. typhimurium* using a DNA probe immobilized on the surface of a quartz crystal [26]. The prospects of microbial detection by genetic sensor were shown also when using optical and electrochemical transducers. So, the process of hybridization of oligonucleotides was registered optically by the method of surface plasmon resonance [27] as well as evanescent waves registration [28]. In the latter case, the sensor allowed the repeated registration of 1 mg of oligonucleotide containing 20 bp in 60 sec.

The procedure of detecting pathogenic microorganisms in drinking water by DNA analysis has been presented. The method is based on sample DNA hybridization with species-specific sequences of 16S rRNA loci; hybridization products are detected using a fluorescent label. With the stages of pre-concentration of a sample and cultivation of cells, it is possible to reach the lower detection limit of about 1 cell/100 ml. The possibility of detection of *Aeromonas hydrophila* and *P. aeruginosa* cells has been demonstrated [29].

A biosensor prototype described quite recently permits the detection of specific DNA sequences and is based on the measurement of values of the peak current generated by hybridization indicators. Indicators were charged cobalt-organic complexes, which had been preliminarily concentrated on the surface of a glass-carbon electrode through binding them by the regions of a two-stranded DNA formed as a result of hybridization [30].

A biosensor for measuring total DNA content of microbial cells has been developed [31]. With this purpose, a biotin-labeled protein of E. coli, which binds single-stranded DNA and the antibodies to nonspecific DNA conjugated with urease, was incubated in the presence of streptavidin with target samples. The resulting complexes were immobilized by filtration on a membrane containing biotin complexes. Urease activity was measured by LAPS. Sensor signals were proportional to DNA concentration in the range of 2–200 pg. Total assay time was no more than 2 h.

Still another approach to microbial detection is based on detection of their enzymatic activities. Such procedures are characterized by the high rate of analysis but allow the detection of only the total content of microorganisms in a sample, with no possibility of species-specific analysis. Examples of the works describing this approach are given below.

The method of non-contact detection of living microbial cells in food stuff has been developed. The method is based on registration of fluorescence of metabolites typical of living microorganisms. The lower detection limit was 20 cells/cm^2 of the surface; assay time was a few seconds [32].

The amperometric method of detection of *E. coli* bacteria using the λ phage has been described. The method is based on determination of the activity of β-galactosidase excreted from cells destroyed as a result of phage lysis. Combination of the developed procedure of analysis with sample concentration and incubation made it possible to detect 1 CFU/ml in 6 h. The developed plant allowed for the concurrent analysis of eight samples [33].

One more approach to *E. coli* detection is based on determination of catalase activity in microbial cells. Peroxide electrode was used as a transducer; the registered parameter was reduction of hydrogen peroxide concentration. The assay time was less than 10 min; the lower detection limit was 10 CFU/200 μl. The efficiency of sorption of the bacteria on a matrix was shown to depend on cell concentration in the solution [34].

Modification of this approach may be based on the concepts of uniqueness of biochemical characteristics expressed by the ability for utilization/oxidation of particular substrates for each bacterial species. This information is the only characteristic of the taxon in question and is not absolutely repeated in other microorganisms. This offers a possibility of developing express methods for microbial detection in samples of different nature obtained from the air, surfaces of various objects, water, and soil. Unfortunately, this approach has also serious restrictions and can be used only in case of the overwhelming prevalence of target microorganisms in a sample.

Up to now, comparatively few publications on this subject have been presented. The typical characteristic of this kind of research is the study of such problems as the activity of different metabolic systems of microorganisms [35]. The promising trend is development of biosensor methods for detection of microorganisms based on registration of their oxide-reductase (including dehydrogenase) activity. The dehydrogenase activity of cells can be analyzed using standard biosensor transducers: oxygen, ion-selective, or mediator electrodes. The receptor element of biosensor in this case will be a carrier containing microorganisms taken from the environment/sample. The obtained analyzer registers the activity of microbial enzyme systems; in some cases, under predominance of any particular species (strain) in a sample, it is probable to identify it by the typical pattern of substrate specificity which is a ratio of activity values of the catalytic systems of cells at introduction of different test substrates. Form the standpoint of bacterial identification, it is practically important to study the patterns of substrate specificity in microorganisms of different genera and species-specific groups, to apply modern theories of multifactor analysis (chemometry, theory of pattern recognition) for selective detection of substrates [36, 37].

In the present work, we have evaluated the possibility of identification of conditionally pathogenic microorganisms by their enzymatic activities, optimized the conditions of functioning of immunosensor based on pH-sensitive field transistor, and presented a model of immunosensor for detection of *Legionella micdadei* cells.

Methods

Materials.

Glucose oxidase from *Aspergillus niger* (ICN Pharmaceuticals, USA), horseradish peroxidase, urease, *o*-phenylenediamine (OPD), ascorbic acid, xylose, arabinose, bovine serum albumin, Tween-20 (all from Sigma, USA), hydrogen peroxide (Khimreaktiv, Russia), non-fat dry milk (Amersham, UK); glucose, ethanol, glycerol, urea, sodium chloride (Diacon, Russia) were used in the experiments. Phosphate buffer solutions (1 mM, pH 6.0) were prepared by mixing standard stock solution of K_2HPO_4 and NaH_2PO_4. Antibodies and peroxidase conjugates were generously provided by the Laboratory of the Bases of Molecular Pathology, Research Institute of Applied Microbiology, Obolensk.

Nitrocellulose membranes Hybond-N, Hybond-N+, Hybond-C (Amersham, UK, pore size 0.45 μm), Millipore (Sigma, USA, pore size 0.22 μm), Vladipore (Polimersyntes, Russia), MFS (Advantec, USA, pore size 0.45 μm), PALL (Pall), and BIO-RAD (Shleicher&Schuell), and chromatographic paper GF/A (Whatman) were used in assays.

The immunoassay was carried out in Costar optically transparent polystyrene microplates.

Transducer.

Field-effect transistors (FETs) with a Ta_2O_5 gate insulator were produced by Avangard Company (Saint-Petersburg, Russia). Their pH-sensitivities were in the range of 50–55 mV/pH. The FET signals were registered at the constant voltage gate mode (Blackburn, 1987). Drain currents I_D were in the range 100–150 μA. Drain voltage U_D was 2.0 V and gate voltage U_G was 0.3 V, applied to reference Ag/AgCl electrode (Reshetilov et al., 1992).

The investigation of enzymatic activities of microorganisms were carried out using standard Clark type electrodes.

Immobilization of Microbial Cells and Analysis on the Basis of Metabolic Tests.

Microorganisms were immobilized by filtration of 10 ml of suspension containing the known quantity of cells through a GF/A or Vladipore fragment of 110 mm^2. The obtained receptor was fixed on the working surface of Clark electrode. Measurements were made in a flow cell of 50 μl; the registered parameter was the maximal rate of change of the electrode signal at substrate oxidation by the cells. Nonpathogenic strains of *L. micdadei*, *S. enteritidis*, *M. bovis*, and spores of the strains *A. niger* and *Stachybotris chartarum* were used as target microorganisms.

Immobilization of Enzymes and Registration of Enzyme Activities.

The 50 μl of enzyme solution was applied to a membrane fragment of 3×3 mm using a microfiltration device and dried in the air for 10 min.

The membrane was fixed on a transistor by special microholder. The transistor was placed into a glass cuvette filled with 2 ml of buffer solution. When registering the horseradish peroxidase activity, 10 μl of ascorbic acid and 10 μl of OPD were added into the cuvette and the base level of the signal was registered. Then the reaction was initiated by addition of 10 μl of hydrogen peroxide. Signal change was registered. In case of glucose oxidase and urease, the base level of the signal was registered in the buffer, then 20 μl of enzyme substrate was added (glucose or urea, respectively), and signal change was registered. The reaction was performed at room temperature under continuous stirring.

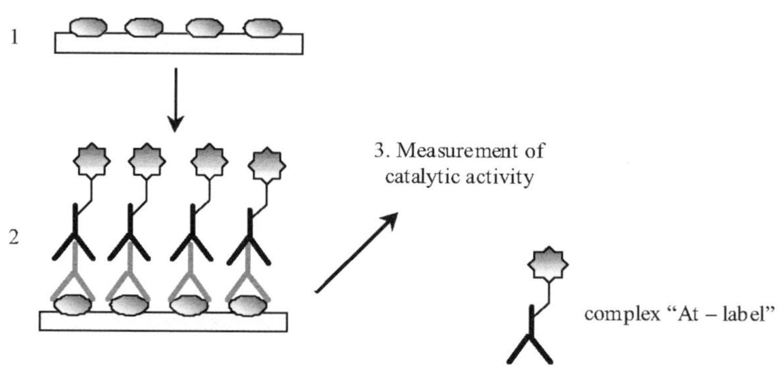

Figure 1. The scheme of immune-enzyme detection of microorganisms

The receptor element of the sensor was formed by the scheme presented in Figure 1, where horseradish peroxidase was used as an enzyme marker.

Immobilization of Cells (step 1, Figure 1).

The 1 μl of suspension of killed L. micdadei cells in buffer solution was applied to a nitrocellulose filter (3×3 mm^2) and dried in the air for 30 min.

Immunoassay Protocol (step 2, Figure 1).

Immunochemical reactions were carried out on immunological plates. Incubation at all stages of the reaction was performed under gentle shaking at 37 °C.

Filters with immobilized microorganisms were placed into holes of a 24-hole plate, and 1 ml of blocking solution was poured in. Incubation was for 1 h. Then the solution was removed and the filter was washed with PBS-T solution (phosphate buffer with 0.05 % Tween-20). The 0.5 ml of solution of monoclonal antibodies in the working dilution (1:100) was introduced into each hole and incubated for 30 min. Then the filters were washed 5–6 times with PBS-T solution. At the next stage, the 0.5 ml of solution of peroxidase conjugate of anti-species antibodies in the working dilution (1:100) was introduced into holes and incubated for 30 min. The filters were washed 5–6 times with PBS-T solution.

Registration of Signals Using FET (step 3, Figure 1).

The prepared receptor element of the sensor was fixed on transistor surface using special microholder. Then the level of enzyme label activity was registered (as mentioned above for determination of the activity of immobilized enzyme).

Results

Detection of Microorganisms on the Basis of Registration of Their Enzymatic Activity

The substrate specificity of microorganisms has been analyzed with the purpose to assess the possibility of their qualitation. The graphs of substrate specificity of strains under study are shown in Figure 2. From the standpoint of practical appropriateness, the number of substrates used for identification of microorganisms was limited to five compounds.

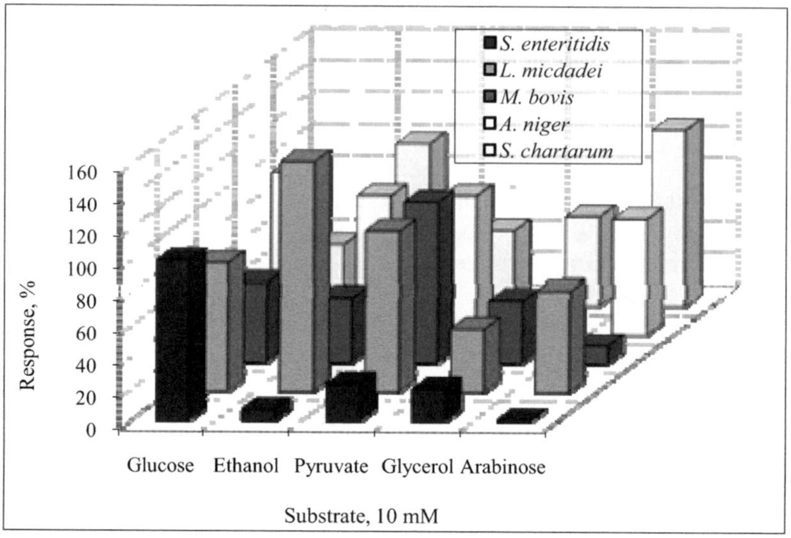

Figure 2. Substrate specificity of target microorganisms

As the diagram shows, the ratio of responses to the given substrates is unique for each strain. Thus, each of the tested strains is characterized by a unique "pattern" of substrate specificity, which allows its qualitative identification in case of its predominance in a sample.

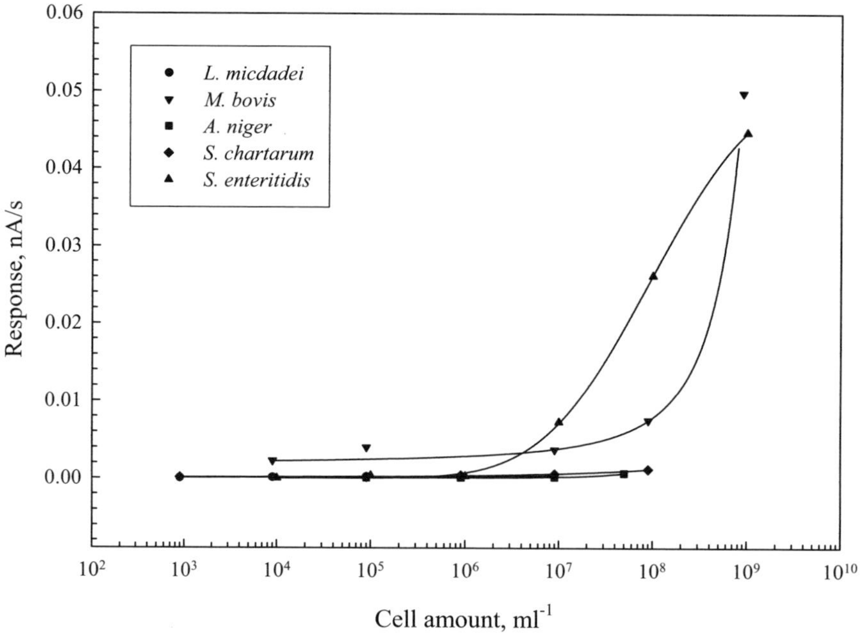

Figure 3. Sensor response dependence on microbial concentrations

The graphs of sensor response dependence on cell concentrations are shown in Figure 3. Glucose was a substrate for *S. enteritidis* and spores of fungal strains; pyruvate was a substrate for *L. micdadei* and *M. bovis*. It can be seen that the absolute value of responses generated by receptors on the basis of *L. micdadei* and spores of fungal cultures is low, so that the calibration curves of these receptors are plotted on a separate graph (Figure 4). The lower detection limit for most strains was about 10^6 cells/ml. The receptor based on *M. bovis* was characterized by sufficiently high signals, but the lower detection limit of its cells was approximately 3×10^7–5×10^7 cells/ml. *S. enteritidis* was characterized by the high value of responses and the lower detection limit of about 10^6 cells, which made it optimal in view of further studies.

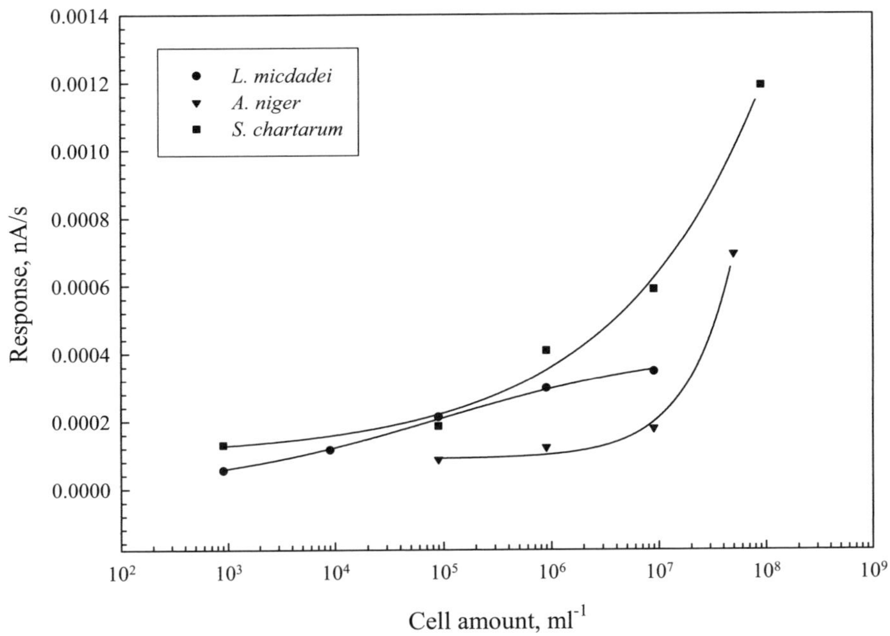

Figure 4. Dependences of sensor response on concentrations of microorganisms *L. micdadei*, *A. niger*, and *S. chartarum*

To estimate the possibility of identification of target microorganisms on the background of nonspecific microflora, the specificity of background microflora samples taken from the premises of Serpukhov poultry factory and a mixture of these samples with *S. enteritidis* cell suspension has been assessed. The results are presented in Figures 5, 6.

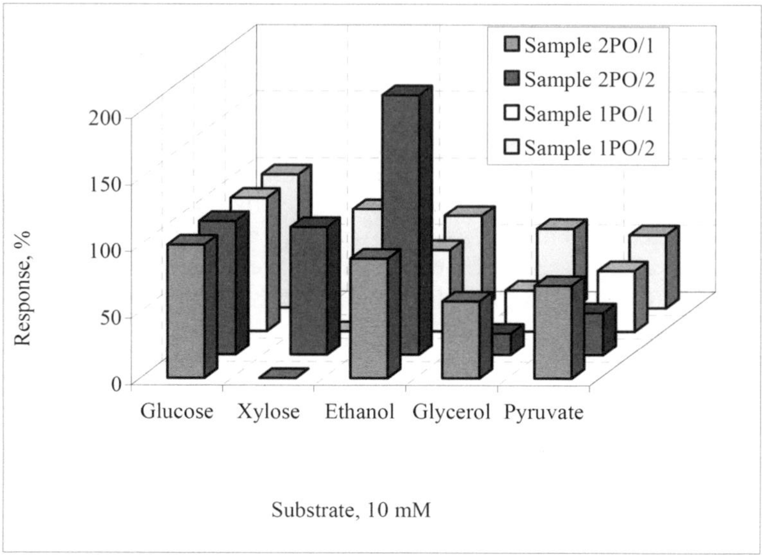

Figure 5. Substrate specificity of background microflora samples

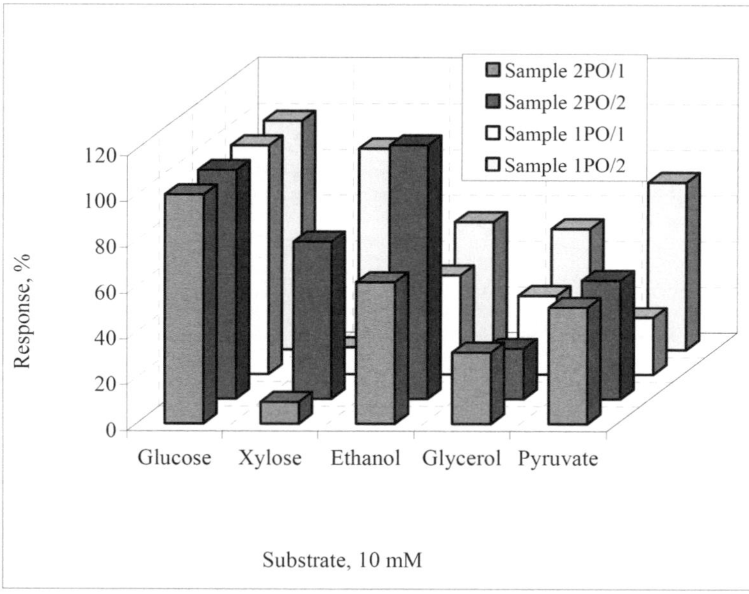

Figure 6. Substrate specificity of the mixture of background microflora samples and *S. enteritidis* cells

Experimental results show that the metabolic patterns of background microflora and the obtained mixture are significantly different: the significance level of their similarity determined by Student is 0.73. The main difference consists in the increase of absolute value of mixture response to glucose, which is the most preferable substrate for *S. enteritidis*.

Thus, the described approach is acceptable for estimation of target cells concentration in pure culture or qualitation of target microorganisms on the background of microflora with the known metabolic pattern. In the presence of a lot of unknown background microorganisms in a sample, this approach may become unpractical.

Immunoassay-Based Microbial Detection

Matrix Type Selection

Physicochemical parameters of membranes (pore size, structure, ion-exchange and, in particular, proton-binding properties) affect the rate of diffusion of the products of substrate oxidation by immobilized enzyme. In turn, the rate of diffusion determines sensor measurement time. Thus, the comparative characterization of diffusion permeability of different membranes (a matrix of sensor receptor element) is extremely important. This value was measured by registration of transistor signals appearing at a discontinuous variation of pH values in the solution. Membrane fragments were placed on the gate region of transistor, followed by addition of 10 μl of 0.1 M HCl (final concentration in the cuvette, 0.5 mM) and registration of parameter $t_{90\%}$ (the time of reaching 90% of the stationary value by the signal) correlating with the rate of proton diffusion in this material. Measurements were made in 1 mM phosphate buffer solution (pH 6.0). The proper response time of transistor (containing no membrane) was negligibly low: no more than a few seconds. The results of measurements are shown in Table 2 ($t_{90\%}$ and initial rate of reaction measured by FET, v_0, are presented).

Table 2. Diffusion properties of membranes

	$t_{90\%}$, s	v_0, mV/s
GF/A	>200	0.27
Millipore	87	1.35
Vladipore	100	1.10
Hybond-N	>260	0.19
Hybond-N+	>260	0.19
Hybond C	112	0.77
PALL	250	0.46
BIO-RAD	230	0.30
MFS	98	0.96

Membranes are arranged in the following series by the increase of diffusion permeability (decrease of $t_{90\%}$): Hybond-N – Hybond-N+ – PALL – BIO-RAD – GF/A – Hybond C – Vladipore– MFS – Millipore. With the Millipore membrane, the maximal signal value was reached rather quickly (the time of reaching 90% of the maximal response was 87 sec). The Hybond membranes proved to be "slower". The period of time from substance introduction to the beginning of response was quite significant (as compared with Millipore), and 90% of the maximal signal value was reached in more than 260 sec after acid introduction. These results indicate that the Millipore, MFS, and Vladipore membranes are more preferable at signal registration by pH-sensitive field-effect transistor; they are characterized by the lowest periods of time $t_{90\%}$ and provide the rapid diffusion of protons to transistor gate region and the high initial rate of response. The above factors are important for reducing the time of assay as well as errors at measurement of the initial rate of enzymatic reaction, the registered value of which may be limited by slow diffusion of reaction products through the membrane. At the same time, only Hybond membranes provided the density of immune components sufficient for immunoassay (for comparison: analytical signals obtained from immune complexes immobilized on Millipore membranes were no more than 10% of the signals registered with Hybond membranes). Hence, we used Hybond membranes in further experiments. Thus, it is advisable to use Millipore and Vladipore membranes for direct detection of corpuscular antigens and Hybond membranes for systems with immobilization of antibodies and competitive binding of low-molecular antigens.

Selection of Enzyme Label

In immune-enzyme assay, the quantity of enzyme label reflects the content of analyte in the system. High specificity and specific catalytic activity of the enzyme permit detection of enzyme label at low concentrations and, as a consequence, low concentrations of analyte. For development of such immunosensor, it is necessary to select an optimal enzyme label. Enzyme is selected with regard to the factors such as specificity, catalytic enzyme activity, availability, stability, etc. Besides, the choice of enzyme label is determined by a system of enzyme activity registration. Enzyme labels most often used in immune-enzyme assay are glucose oxidase (GOD), horseradish peroxidase (HRP), and urease [38-42]. In our work, we have compared these three enzymes using FET. The enzymes were immobilized on Hybond-C membranes. Table 3 shows the results obtained: linear ranges of determined concentrations and lower detection limits.

Table 3. Enzyme characteristics

Enzyme label	Lower detection limit, ng/mm^2	Linear range of concentrations
HRP	0.001	0.002–30 ng/mm^2
GOD	5	0.700–100 μg/mm^2
Urease	1	5–6700 ng/mm^2

The table shows that HRP is the most acceptable enzyme label in immune-enzyme assay with application of electrochemical detection. The lower limit of its detection is 0.001 ng/mm^2: much less than with GOD and urease. In the course of biosensor development, this property will make it possible to increase assay sensitivity. HRP provides the maximal change of pH during the reaction and the maximal initial reaction rate, which is necessary for assay time reduction. Moreover, to obtain a major pH shift in case of HRP it is necessary to introduce 2–3 substrates into the reaction medium, which inevitably increases the measurement error. Urease is a single-substrate enzyme and hence more preferable for immunoassay. At the same time, it is slightly inferior to HRP by its characteristics (the minimal detectable concentration is 1 ng/mm^2, whereas for HRP it is 0.001 ng/mm^2). Therefore, we have used HRP as enzyme label when developing the model of a sensor.

Analysis of Enzyme Substrates

Express detection of HRP requires the measurement of parameters of quick responses. Such a parameter may be concentration of hydrogen ions. There is a wide range of reducing agents, peroxidase oxidation of which is accompanied by pH change in the solution. Nevertheless, it is necessary to select substrates and reaction conditions, at which pH change will be sufficiently quick and proportional to HRP concentration. In our early experiments it was shown that the most effective pH shift as a result of peroxidase reaction occurs with the following substrate mixture: OPD, ascorbic acid, and hydrogen peroxide. The screening of different phenols and anilines as co-oxidizing compounds indicated that the combination of OPD, ascorbic acid and hydrogen peroxide allows reaching the maximal substrate-substrate activation [43]. The OPD plays the role of electron mediator in the transformation of ascorbate to hydroascorbate. Later on [44] we showed that the increase of concentrations in the range of 10^{-4}–10^{-3} M for all substrate components results in a higher sensor signal. For the range of 10^{-3}–5×10^{-3} M, the increase of ascorbic acid and OPD concentrations leads to a signal decrease. At the same time, the increase of H_2O_2 concentrations correlates with further increase of sensor signal. The obtained dependences made it possible to determine the optimal substrate mixture for further studies: OPD, 1 mM; ascorbic acid, 1 mM; hydrogen peroxide, 3 mM.

Characteristics of the Developed Immunoassay

A model of immunosensor for detection of *L. micdadei* cells has been developed taking into consideration the developed optimal conditions of the functioning of FET-based sensor. The scheme of analysis is shown in Figure 1. The 5% non-fat milk solution was used as a blocking reagent, with the addition of 1 ml of solution into each hole after immobilization of microbial cells. A 2-fold decrease of the amount of blocking solution increases the nonspecific binding, which, in its turn, enhances the measurement error to 12–19%. Further increase of the amount of blocking reagent has no effect on experimental results. With 3% solution of bovine serum albumin (BSA) used as a blocking agent, it was possible to increase sensor signals 5–10-fold but the measurement error increased as well (> 20%). Thus, the calibration curve was plotted using 5% non-fat milk solution, with the addition of 1 ml of solution into each hole in the course of assay.

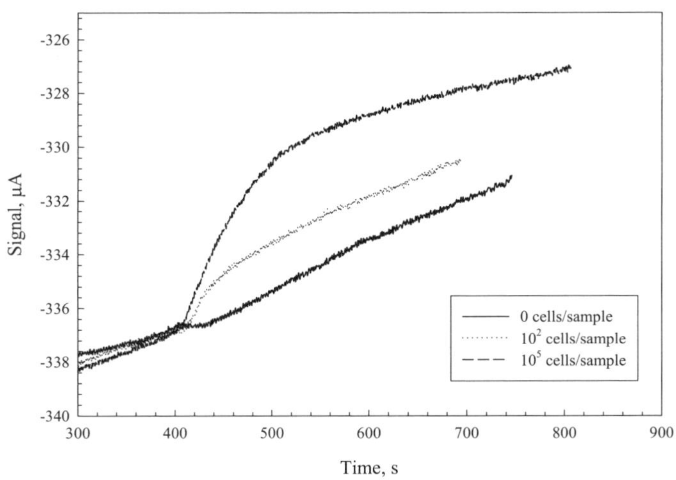

Figure 7. Types of FET signals

The typical kinds of FET signal dependences on time are shown in Figure 7.

Figure 8 shows the calibration dependence of immunosensor signals on the concentration of microbial cells.

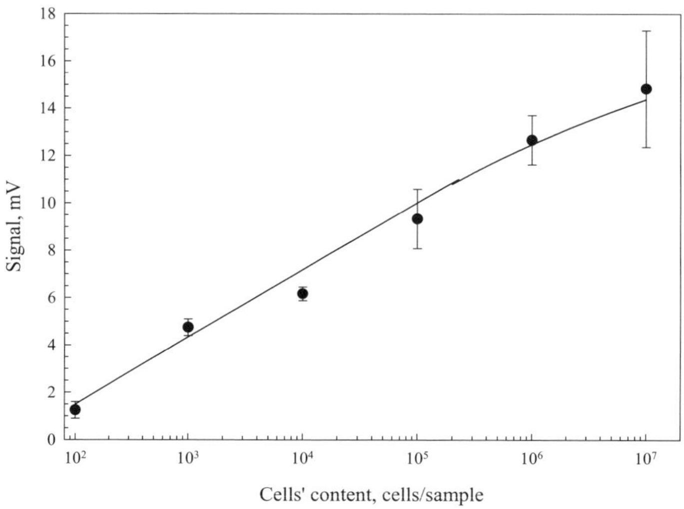

Figure 8. Calibration dependence for detection of microbial cells

After the analysis, the used membrane was removed and replaced by a new one to avoid regeneration of transistor surface, as it would be necessary at the formation of immune complexes directly on transistor surface [39, 45, 46].

Thus, the developed immunosensor model can be used for detection of cells with the lower limit of 10^2 cells, which indicates a rather high sensitivity of the developed method. At the same time, the coefficient of variation was no more than 7%. Tested sample volume was 1 μl without preliminary enrichment; assay time was no more than 3 h. Application of simple instruments offers a possibility to perform the analysis directly in the place of sample taking.

Conclusions

Thus, the model of the immunosensor created in the framework of the present study can carry out reliable determination of the target cells with the lower limit of detection of 10^2 cells/sample within 3 h. The sensor based on the registration of enzymatic activity of the target cells is able to detect 10^6 cells/sample and ensures the recognition of the detected microorganism only in the case of pure culture or qualitation of target on the background of microflora with the known metabolic pattern. So it can be used only for rapid preliminary analysis with following more reliable assay enabling immune or genetic analysis techniques.

Acknowledgements

The study was supported partially by grant 2.1.1.7789 of Analytical Departmental

Program «Development of scientific potential of the higher school (2006-2008)» in 2006-2007, subdivision № 2.1.1 «Fundamental research in the field of natural sciences». Authors also gratefully acknowledge Dr. S.F. Biketov and Dr. E.V. Baranova (Research Institute of Applied Microbiology, Obolensk) for provided immunoreagents and consultations on immunoassay

References

[1] D. Ivnitski et al., Biosensors for detection of pathogenic bacteria, *Biosensors & Bioelectronics* **14** (1999), 599-624.
[2] R.H. Hall, Biosensor technologies for detecting microbiological foodborne hazards, *Microbes Infect* **4** (2002), 425-32.
[3] S. Varfolomeev et al., Chemical and biological safety. Biosensors and nanotechnological methods for the detection and monitoring of chemical and biological agents, *Pure & Applied Chemistry* **74** (2002), 2311-2316.
[4] A. Rasooly et al., Biosensors for the analysis of food- and waterborne pathogens and their toxins, *J AOAC Int* **89** (2006), 873-83.
[5] A.A. Bergwerff et al., Surface plasmon resonance biosensors for detection of pathogenic microorganisms: strategies to secure food and environmental safety, *J AOAC Int* **89** (2006), 826-31.
[6] M. Diaz-Gonzalez et al., Immunosensor for Mycobacterium tuberculosis on screen-printed carbon electrodes, *Biosens Bioelectron* **20** (2005), 2035-43.
[7] F. He et al., Rapid diagnosis of *M. tuberculosis* using a piezoelectric immunosensor, *Anal Sci* **18** (2002), 397-401.
[8] T. Geng et al., Detection of low levels of *Listeria monocytogenes* cells by using a fiber-optic immunosensor, *Appl Environ Microbiol* **70** (2004), 6138-46.
[9] I. Abdel-Hamid et al., Flow-through immunofiltration assay system for rapid detection of E. coli O157:H7, *Biosens Bioelectron* **14** (1999), 309-16.
[10] M. Yemini et al., Specific electrochemical phage sensing for *Bacillus cereus* and *Mycobacterium smegmatis*, *Bioelectrochemistry* **70** (2007), 180-4.
[11] J.M. Libby et al., Detection of *Neisseria meningitidis* and *Yersinia pestis* with a novel silicon-based sensor, *Journal of Clinical Microbiology* **27** (1989), 1456-1459.
[12] B. Konig et al., Detection of viruses and bacteria with piezoelectric immunosensors, *Analytical Letters* **26** (1993), 1567-1585.
[13] L.K. Cao et al., Detection of *Yersinia pestis* fraction 1 antigen with a fiber optic biosensor, *J Clin Microbiol* **33** (1995), 336-41.
[14] W.E. Lee et al., Rapid detection and identification of biological and chemical agents by immunoassay, gene probe assay and enzyme inhibition using a silicon-based biosensor, *Biosens Bioelectron* **14** (2000), 795-804.
[15] W.E. Lee et al., Rapid immunofiltration assay of Newcastle disease virus using a silicon sensor, *J Immunol Methods* **166** (1993), 123-131.
[16] E. Prusak-Sochaczewski et al., Development of a piezoelectric immunosensor for the detection of Salmonella typhimurium, *Enzyme Microb Technol* **12** (1990), 173-177.
[17] N.F. Starodub et al., Construction and biomedical application of immunosensors based on fiber optics and enhanced chemiluminescence, *Opt Eng* **33** (1994), 2958-2963.
[18] V.K. Akimenko et al., A rapid method for detection of Clostridium thermocellum by field-effect transistor-based immunodetection, *Journal of Microbiological Methods* **24** (1996), 203-209.
[19] A.Yu. Karulin et al., Polyvalent antigen-antibody interaction: theoretical models and role in immunoassay of cells, *Journal of the All-Union Mendeleyev Chemical Association* **34** (1989), 30-38.
[20] H.-Y. Mason et al., Detection and identification of microbial contamination, Proceedings of *7th World Congress on Biosensors*, Kioto, 2002, C2.01.
[21] I. Abdel-Hamid et al., Portable air-quality monitoring system, Proceedings of *7th World Congress on Biosensors*, Kioto, 2002, P1-2.02.
[22] Z. Muhammad-Tahir et al., A conductometric biosensor for biosecurity, *Biosens Bioelectron* **18** (2003), 813-819.
[23] T. Matsuura et al., The specific detection of Salmonella by using the dissolved oxygen measurement device, Proceedings of *7th World Congress on Biosensors*, Kioto, 2002, P1-2.30.
[24] Y. Li et al., Capillary Bioseparator/bioreactor-based Electrochemical Biosensor for Detection of Escherichia coli O157:H7 in Food Samples, Proceedings of *7th World Congress on Biosensors*, Kioto, 2002, P1-2.54.

[25] S.-I. Tu et al., Simultaneous Detection of Escherichia coli O157:H7 and Salmonella Typhimurium in Foods Using Immunomagnetic Capture and Lanthanide Time-Resolved Fluorescence, Proceedings of *7th World Congress on Biosensors*, Kioto, 2002, P1-2.57.
[26] M.E. Downs, Prospects for nucleic acid biosensors, *Biochem Soc Trans* **19** (1991), 39-43.
[27] D. Pollard-Knight et al., Immunoassays and nucleic acid detection with a biosensor based on surface plasmon resonance, *Ann Biol Clin* **48** (1990), 642-646.
[28] C.R. Graham et al., Gene probe assays on a fibre-optic evanescent wave biosensor, *Biosens Bioelectron* **7** (1992), 487-493.
[29] C. Jorgensen et al., Development of a DNA-chip for the detection of pathogens in drinking water, Proceedings of *7th World Congress on Biosensors*, Kioto, 2002, P1-1.21.
[30] K.M. Millan et al., Sequence-selective biosensor for DNA based on electroactive hybridization indicators, *Anal Chem* **65** (1993), 2317-2323.
[31] V.T. Kung et al., Picogram quantitation of total DNA using DNA-binding proteins in a silicon sensor-based system, *Analytical Biochemistry* **187** (1990), 220-227.
[32] F. Cleary et al., Detection of Microbial Contamination on Foods, Proceedings of *7th World Congress on Biosensors*, Kioto, 2002, P3-4.09.
[33] A.S. Mittelmann et al., A phage-based electrochemical sensor for the specific detection of bacteria, Proceedings of *7th World Congress on Biosensors*, Kioto, 2002, C1.06.
[34] N. Sippy et al., Rapid electrochemical detection and identification of catalase positive micro-organisms, Proceedings of *7th World Congress on Biosensors*, Kioto, 2002, A3.11.
[35] K.A. Lusta et al., Physiological and biochemical peculiarities of Gluconobacter oxydans and prospects of its application in biotechnology and biosensor systems (review), *Applied Biochemistry and Microbiology* **34** (1998), 339-353.
[36] A.N. Reshetilov et al., Elements of the theory of pattern recognition in detection of ethanol content in mixture using microbial and enzyme sensors, *Sensor Systems* **12** (1998), 486-496.
[37] A.N. Reshetilov et al., Detection of ethanol in a two-component glucose/ethanol mixture using a nonselective microbial sensor and a glucose enzyme electrode, *Biosens Bioelectron* **13** (1998), 787-93.
[38] H. Sakai et al., Availability and development of an enzyme immunomicrosensor based on an ISFET for human immunoglobulins, *Analytica Chimica Acta* **230** (1990), 189-193.
[39] C. Colapicchioni et al., Immunoenzymatic assay using CHEMFET devices, *Sensors and Actuators B* **4** (1991), 245-250.
[40] K. Dill et al., Detection of human asialo-alpha(1)-acid glycoprotein using a heterosandwich immunoassay in conjunction with the light addressable potentiometric sensor, *Glycoconj J* **13** (1996), 637-641.
[41] A.G. Gehring et al., Use of a light-addressable potentiometric sensor for the detection of Escherichia coli O157:H7, *Anal Biochem* **258** (1998), 293-298.
[42] A.L. Ghindilis et al., Immunosensors: electrochemical sensing and other engineering approaches, *Biosens Bioelectron* **13** (1998), 113-131.
[43] S.M. Khomutov et al., The study of efficiency of some substrates for pH detection of peroxidase activity, *Journal of Analytical Chemistry* **50** (1995), 659-662.
[44] Y.V. Plekhanova et al., A new assay format for electrochemical immunosensors: polyelectrolyte-based separation on membrane carriers combined with detection of peroxidase activity by pH-sensitive field-effect transistor, *Biosens Bioelectron* **19** (2003), 109-114.
[45] M. Gotoh et al., Immuno-FET sensor, *J Mol Catal* **53** (1989), 285-292.
[46] N.F. Starodub et al., Immunosensor for the determination of the herbicide simazine based on an ion-selective field-effect transistor, *Analytica Chimica Acta* **424** (2000), 37-43.

L.M. Lechuga et al. (Eds.)
IOS Press, 2008

Molecular identification through membrane-engineering (MIME): state-of-the-art biosensor technology for instant, ultra-specific and ultra-sensitive detection of infectious disease agents at global scale

Spiridon KINTZIOS
EMBIO
Faculty of Agricultural Biotechnology, Agricultural University of Athens

Abstract. In recent years there has been a rapid increase in the number of diagnostic applications based on biosensors, including live, intact cells, tissues, organs or whole organisms. In similar fashion to DNA and protein microarrays, which deliver multiplex detection via the high-density spatial arrangement of molecular recognition elements, arrays of cells at high-density can form the basis of cell-based sensors with extremely high-throughput capability. The expression of receptors of interest within these arrays could yield cell-based sensors with defined specificities. In addition, transfected cell microarrays composed of high-density arrays of mammalian cells expressing de-fined genes, could be the basis for future high-throughput cell-based protein sensing platforms.

The artificial insertion of receptor-like molecules in the cell membrane is an attractive alternative to cell transformation with genes expressing membrane-bound antibodies. This generic approach is called **M**olecular **I**dentification through **M**embrane **E**ngineering (MIME). Interaction of MIME cells with viral particles can trigger changes to the cell membrane potential that are measured by appropriate microelectrodes, according to the principle of the Bioelectric Recognition Assay (BERA). BERA is a biosensory method based on a unique combination of a group of cells, whose immobilization in the matrix preserves their physiological functions and measures the expression of the cell interaction with viruses, through the change in electrical properties. In this way, when a positive sample is added to the probe, a characteristic, 'signature-like' change in electrical potential occurs upon contact between the virus and the gel matrix. BERA has been used for the detection of viruses in humans (Hepatitis B and C viruses, herpes viruses), animals (prion protein, foot and mouth disease, blue tongue virus) and plants (tobacco and cucumber vi-ruses) in a remarkably specific, rapid (1-2 minutes), reproducible and cost-efficient fashion. The sensitivity of the virus detection with BERA is equal or even better than with advanced immunological, cytological and molecular techniques, such as the reverse transcription polymerase chain reaction (RT-PCR).

The BERA biosensor diagnostic system is currently available as a desktop, laboratory-scale prototype that can be operated by both expert and lay users. The commercialization process of the device includes engineering for a more compact, stand-alone unit. The system comprises a consumable miniature biosensor (with integrated circuitry, an immobilization matrix and virus-specifically responding cells), a data acquisition system and a PC (desktop or laptop). One of the major advantages of BERA is the extended storability of the disposable sensors, which is also documented by other research groups. So far, more than 35000 sensors have been used for screening worldwide.

Since BERA measurements are essentially electric signals, they can be instantly evaluated by means of specific software either on site (stand alone devices) or via an Internet site. Desktop devices with Internet-based evaluation are targeted to small clinic and doctors' office screening tests in the USA and European Union. On site assays are related to portable BERA field test kits, which are ideal for clinical testing in developing countries and military applications. In this way, BERA and MIME cell sensors lay the foundation for a fully operational global biothreat monitoring network.

Introduction: the new era of cellular biosensors

A biosensor is a device that detects, transmits and records information regarding a physiological or biochemical change. In recent years there has been a rapid increase in the number of diagnostic applications based on biosensors, including live, intact cells and -in some cases- tissues, organs or whole organisms. Whole cells provide multipurpose catalysts, particularly in processes that require the participation of a number of enzymes in sequence; therefore, the utilization of whole cells as a source of intracellular enzymes is often a better alternative to purified enzymes in various industrial processes. [1,2]. Cellular biosensors have the potential to become the next revolution in diagnostic technology, following immunoassays and nucleic acid technologies. Contrary to any other system, a cell biosensor can provide information on the infectious properties of novel viruses or viral strains. This is particularly important when no other detection method is available for this novel virus. This interesting property lies in the inherent similarity of cell biosensors with cell culture, namely the dual character of a detection system that is also a host of the pathogen under detection. The culture of cells for detecting and even determining the presence of a pathogen is considered as the golden standard of the diagnostic community. However, several weeks may be required before a clearly defined symptom is observed in a test culture [3-4]. On the contrary, cell sensors required significantly less time (from a few moments to a day) in order to produce a reliable result. This is very important because, the ability to detect a single type of molecule is not necessarily associated with the ability to evaluate the virulence of a pathogen. For example, viral strains expressing novel antigens may escape detection by the most advanced immunoanalytical systems. Nevertheless, they will be detected by an appropriate cell biosensor, provided that the cells used in the analytical system are susceptible (or responding otherwise) to the virus under detection [5].

In order to be able to use a biosensor system for detecting viruses and other pathogens in the field, a number of requirements have to be satisfied. Compliance with these characteristics should be obligatory for claiming the operative appropriateness for any new detection system [6]. It is always desirable to achieve a high assay speed, reproducibility, accuracy, selectivity and sensitivity, as well as extended sensor storability under simple conditions. In addition, the cost of each assay should be kept minimum, along with a minimal sample volume and minimal sample preparation time. In order to have a truly portable, field-applicable system, the use of reagents should also be minimized (especially if reagents require specific measures of disposal) and the sensor must provide as much information as possible during a single assay, with as little time delay as possible. Finally, the biosensor system should be applicable for multianalyte detection (Table 1).

The problem of non-selective response

As a result of relentless trial-and-error testing through countless millennia, cells respond in a rather precise and reproducible, albeit *common* way, against an amazingly large number of different molecules. In other words, cell sensors can exhibit a very poor selectivity. This is a very common problem in toxicity assays, where cell sensors successfully *detect* the presence of a toxic (or genotoxic) compound, but generally *fail to determine* the exact nature of the toxic analyte [7-9].

In recent years a number of cell transfection methods have been developed for increasing cell specificity, with considerable success. Cell transfection methods are based on the cellular uptake of foreign DNA by, mainly, human B cells, therefore leading to the expression of membrane-bound receptors [10-13]. A quite renowned application is the CANARY (Cellular Analysis and Notification of Antigen Risks and Yields) system developed by MIT Lincoln Lab. CANARY has been used for the detection of *Yersinia pestis* and other pathogens [14]. However, the applicability of cellular transfection is limited by the lack of stability and the frequent, unwanted alteration of cellular phenotype.

Table 1: Desired attributes for novel analytical/medical diagnostic technologies[1]

High level of	Low level of	Additional attributes
• Speed	• Cost	• Universal applicability
• Reproducibility	• Sample volume	• Portability
• Accuracy	• Sample preparation	• Reagentless assays
• Selectivity		• Real-time monitoring
• Sensitivity		• Maximum information during a single assay
• Storability		

1 adapted from [6]

Breakthrough technologies: Molecular Identification through Membrane-Engineering and the Bioelectric Recognition Assay (BERA)

Molecular Identification through Membrane Engineering (MIME) is a generic methodology for artificially inserting (usually by electroinsertion) tens of thousands of receptor molecules on the cell surface, thus rendering the cell a very specific responder against analytes binding to the inserted receptors [15]. Receptor molecules can vary from antibodies to enzymes to polysaccharides. The working assumption of the method is that attachment of the target molecule to its respective receptor causes a change in the cell membrane structure, which is measurable as a change in the cell membrane potential.

The **Bioelectric Recognition Assay (BERA)** [16-18] is a biosensory method based on a unique combination of a group of cells, whose immobilization in a matrix preserves their physiological functions and measures the expression of the cell interaction with viruses, through the change in electrical properties. A BERA sensor consists of an electroconductive probe containing components of immobilized cells in a gel matrix. Cells are selected to specifically interact with the virus under detection. In this way, when a positive sample is added to the probe, a characteristic, 'signature-like' change in electrical potential occurs upon contact between the virus and the gel matrix. BERA has been used for the detection of viruses in humans (Hepatitis B and C viruses, herpes viruses) and plants (tobacco and cucumber viruses) in a remarkably specific, rapid (1-2 minutes), reproducible and cost-efficient fashion. The sensitivity of the virus detection with BERA is equal or even better than with advanced immunological, cytological and molecular techniques, such as the reverse transcription polymerase chain reaction (RT-PCR) (Table 2). The main advantages of the system are:

- Ultra-high speed: an assay can be completed in less than 100s.
- High sensitivity: at least 2000 times higher than conventional immunological methods.
- Low cost: not exceeding 10 US$ per assay batch.

BERA sensors based on MIME cells have been successfully applied for the instant detection of HBV, HCV, superoxide, prions, foot-and-mouth disease virus (FMDV) and blue-tongue virus (BTV) [6, 19]. An example showing the immediate detection of prion protein is illustrated in Figure 1. The combined method has also been used for the detection of environmental toxins, such as herbicides [20].

Table 2: Performance of BERA/MIME sensors compared with conventional methods of pathogen identification.

	BERA-MIME CELLS	**ELISA**	**PCR**
Tests for	Virus itself	Antibody	DNA/RNA
Time to obtain results	6-40 seconds	20+ minutes	2+ hours
Sensitivity (detection limit)	<10 pg/ml (2000 times higher than ELISA)	20 ng/ml	1 pg/ml
Specificity	Depends on receptor	Relative	Absolute if strict assay conditions are satisfied
Reproducibility	100%	50-85%	100%
Laboratory infrastructure & training	None	Moderate	Very high
Per test cost	$10+	$30+	$200+
Equipment cost	$1,000+	$7,000+	$50,000+

Competing technologies are primarily based on technology that employs either molecular biological methods or immunological methods. In general, the molecular biological method kits are extremely sensitive, but are relatively expensive and require multiple sample preparation steps and the use of sophisticated equipment and trained personnel. These kits are therefore typically sold to hospitals, clinics, central laboratories, and research laboratories. The immunological methods kits are less sensitive and not very reliable, but inexpensive and simple to use. Consequently, the immunological method kits are generally sold as screening tests, field test kits, doctors' office tests, over the counter test kits, etc.

BERA would be more sensitive than current immunological tests, but competitive in cost, simplicity of use and response time. Moreover, although molecular biology diagnostic kits are becoming faster, simpler and less expensive, BERA should still significantly outperform molecular biology kits in these areas. BERA should therefore be preferable to any alternative technology in almost all current screening applications. Similar to immunological assays, BERA assays the interaction between cell membrane components (such as receptors) and viral proteins. Consequently, its specificity has been largely increased by the utilization of both specific cells and antibodies able to interfere with virus attachment to the cell membrane [21]. However, BERA sensors cannot differentiate between different genotypes of a single virus if differences are restricted on a DNA or RNA level only (i.e. are not expressed on a protein level).

BERA can contribute to an improved selectivity compared to enzymic and immunochemical biosensors. Due to its high sensitivity (a common feature of cellular signal transduction systems) and simplicity, the assay offers an attractive complement for virus detection to molecular techniques, such as RT-PCR. These methods are indeed very sensitive, but fairly expensive (therefore inappropriate for mass sample screening) and require separate laboratory space and equipment in order to avoid sample contamination with exogenous viral DNA, while discrepancies may occur when patient HCV viral loads are monitored using different types of assays [22]. In this context, BERA could be particularly useful in drawing a first, rapid conclusion in the presence of a given virus in an unknown sample, therefore providing an effective sample "filter" prior to the employment of other analytical techniques. Thus, BERA could replace ELISA as a standard diagnostic tool.

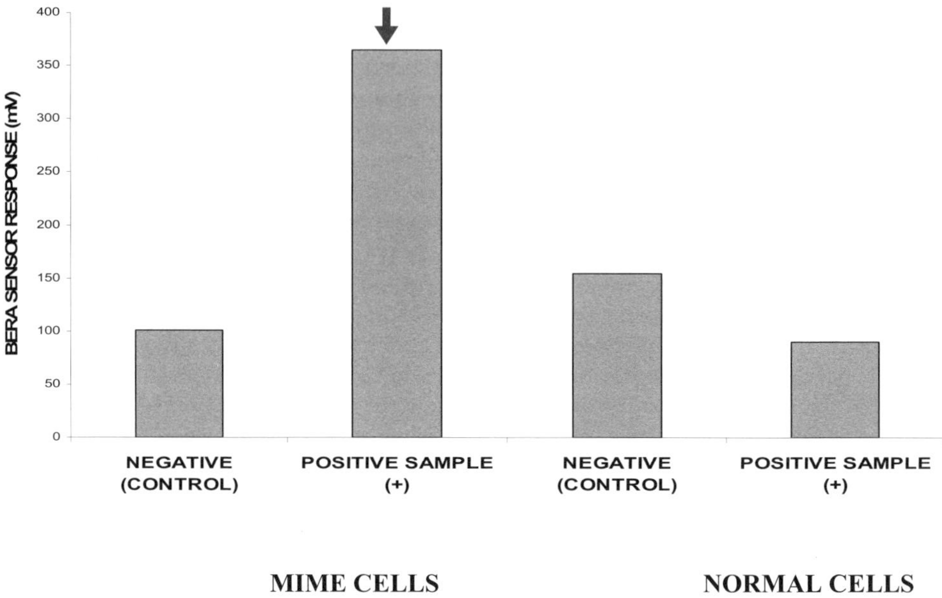

Figure 1. Demonstration of the highly selective response (arrow) of MIME cells bearing receptors for pathogenic prion protein (PrPSc) against a positive, infected sample, as compared with the response of normal, non-engineered Vero cells.

Systems integration

The BERA/MIME diagnostic system is currently available as a desktop, laboratory scale prototype that can be operated by both expert and lay users. The commercialization process of the device includes engineering for a more compact, stand-alone unit. The system comprises a consumable miniature biosensor (with integrated circuitry, an immobilization matrix and virus-specifically responding cells), a data acquisition system and a PC (desktop or laptop). The consumable sensor unit is a very small gel bead containing 40,000-50,000 cells, which is connected to a special ELISA plate bearing pre-fabricated electrodes. This, in turn, is connected to a data logger and finally to a standard PC or laptop. Two different types of sensors are used for different applications (detection assays or pharmacology screening) [23]. The latest (6th) generation of BERA sensors are extremely miniaturized, consisting of a disposable array of gel beads loaded with cells. They are characterized by a very high degree of reproducibility (>99.9%), extremely low cost and speed of manufacturing (with a production performance of approx. 1000 sensors per technician per hour). In addition, the duration of the assay has been reduced from approx. 100 seconds to a mere 40 seconds.

Recently, a significant step was realized towards the automation of the system through the development of a neural classifier, in particular a Multi-Layered Perceptron (MLP), to predict the presence of a virus. An Artificial Neural Network (ANN) was applied with different architecture so as to develop an intelligent system using biosensors for the detection of viruses. The classification stability of the method and the generalization performance, was increased by using a combination of neural classifiers. In this sense, a set of correlated specialized classifiers was produced, which attack the classification problem by applying an appropriate decision combination [24].

Validation of the sensor

In 2002, BERA sensors were field-tested for the first time outside the inventor's laboratory for the diagnosis of HCV at the Kaplan Medical Center (Rehovoth, Israel). In the same year, sensors were constructed and applied for the detection of Venezuelan Encephilitis Virus (VEE), a potent biological warfare agent, at the U.S. Army Medical Research Institute of Infectious Diseases (USAMRIID). In 2003, BERA was included in the North-Atlantic Treat Organization (NATO) Science Program (Project Code: SSTCLG979883) for the detection of infectious agents and toxins.

So far, BERA sensors have been used for the detection of a number of medicinally important viruses, such as HBV, HCV, HSV-1, VZV, CMV, EBV and VEE. More than 50,000 sensors have been used for the detection of HBV and HCV, representing approx. 300 samples. All experiments have been conducted under very strict clinical conditions with appropriate controls. Validation centers included four hospitals, three universities and four military installations.

Commercialization efforts

Since the initial conception of the BERA/MIME technology, a lot of effort has been dedicated to the commercialization of the technology and/or independent products derived thereof. Worldwide patenting of various components and aspects of this novel technology contributed considerably to attracting investors. The first effort was realized by the establishment (in 2002) of BERABiosensors Inc., a New York-based company that carried out initial contacts for the validation of the technology in various US institutions. This was subsequently transformed (in 2004) into BERADiagnostics Inc (Harrison, NY). BERADiagnostics completed a second round of contacts with validation centers in the United States. Dissemination of the method and increase of public awareness about the potential of biosensors for medical diagnostics was fostered by the publication (by BERADiagnostics) of a User's Guide describing, in great detail, the methodological principles, system's operation and interpretation of the results. This process was supported by the availability of instruction videos on the company's website. Recently, a new company, EMBIODiagnostics, has been established using seed funding provided by the Cyprus government for the purpose of further developing the BERA technology. BERA biosensors are also marketed by BioDiagnostix, a UK-based biotechnology company, which plans the scale-up production of the sensors in Europe and elsewhere.

As with other biosensor technologies, BERA/MIME sensors had to satisfy certain market requirements in order to gain wide acceptance by the community of target end-users. There were particular obstacles to be overcome, including the following:

1. A general doubt about the reliability of cell biosensors, in particular their selective response against desired analytes and their satisfactory storability under non-sophisticated conditions. Both issues were effectively resolved

with the invention of MIME technology and the improvement of the cell immobilization procedure.

2. Although the biosensor was producing electric signals as the type of output measurement, clinical microbiologists were strictly accustomed to the "ELISA-plate" assay format. Consequently, the biosensor system was radically redesigned in order to fit this specification: special ELISA plate bearing pre-fabricated electrodes were manufactured and the electronic interface was redesigned.
3. It was important to increase public awareness on the merits of bioelectric measurements, while the method of choice is fluorescence or luminescence. Fluorescence-based systems are less practical than electrophysiological ones, since they are not suitable for quantitative determinations. They are also far more expensive than electrode-based techniques [25]. Fluorescent dyes may also have considerable side-effects on cellular physiology, such as uncoupling of respiratory inhibition and membrane permeabilization [26, 27] and may interfere with the cell parameter that they assay [28].
4. It was equally important to separate market segments requiring a desktop diagnostic system (e.g. clinical microbiology) from those demanding more compact, preferably hand-held devices (e.g. biodefense applications). Considerably more resources are required for the development of the latter type of product, due to the miniaturization of all system components.

Global implementation of BERA/MIME sensors and the future of early warning biodefense systems

Since BERA/MIME measurements are essentially electric signals, they can be instantly evaluated by means of specific software either on site (stand alone devices) or via an Internet site. Desktop devised with Internet-based evaluation are targeted for airports, military installations, small clinic and doctors' office screening tests in the USA and European Union. On-site assays are related to portable field test kits, which are ideal for clinical testing in developing countries and military applications.

The ultimate goal of the war against bioterrorism is the ability to detect pathogen infection incidences at any time, and any place, at a global scale of surveillance. Indeed, only a global early warning system would mobilize the required resources in order to prevent the outbreak of a new pandemic [29]. In this context, we are now in the process of building a universal sensor platform that is based upon a vast network of BERA/MIME sensors. The main features of this network are as follows:

1. Large numbers of small, low power sensors are distributed randomly in specific areas.
2. A sensor network is composed of a large number of sensor nodes, which are densely deployed either inside a phenomenon (e.g. disease outbreak) or very close to it.
3. BERA sensor nodes are intelligent and have the ability to monitor their status and to report their assay results.
4. Topology is built in real time.
5. Periodic updates include the replacement of used or damaged sensors with new ones.

6. The communication of the sensor network with a central site for further evaluation of the remote response is based on either a wired or a wireless mode. In the latter case, the wireless (Bluetooth, WiFi,etc.) transmitter is turned on when: (a) the sensor's detection is positive, (b) a request for assay data transmission is received or (c) all the biosensors are deactivated and need replacement.
7. Within the network, each sensor node can combine information derived from BERA/MIME sensors with signals retrieved from other sensor systems, as presented in Figure 2.

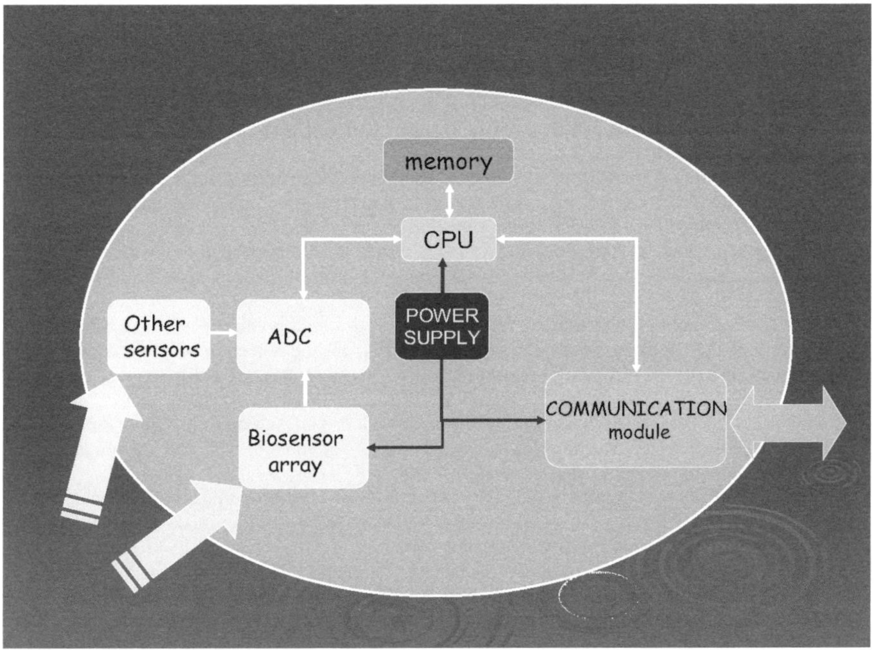

Figure 2. Schematic outline of a networked BERA/MIME sensor node.

References

[1] G.F. Bickerstaff, Immobilization of Enzymes and Cells, Humana Press, Totowa, NJ,1997.
[2] S.F. D'Souza, Microbial biosensors. *Biosens.Bioelectron.* **16** (2001) 373-353.
[3] J.C. Park, Y.S.Hwang, H. Suh, Viability evaluation of engineered tissues.*Yonsei Med. J.* **41** (2000) 836-844.
[4] P. A. Jones, V.A. Baker, A.J.E. Irwin, L.K. Earl, Interpretation of the in vitro proliferation response of MCF-7 cells to potential oestrogens and non-oestrogenic substances. *Toxicol. in Vitro* **12** (1998), 373-382.
[5] J. Odum, S. Tittensor, J. Ashby, Limitations of the MCF-7 cell proliferation assay for detecting xenobiotic oestrogens. *Toxicol. in Vitro* **12** (1998), 273-282.
[6] S. Kintzios, Next generation cell biosensors for field-based virus detection. *Proc. 6th Biodetection Conference*, Knowledge Foundation: Washington, DC, 2006.
[7] K.L.E.Kaiser, V.S. Palabrica, *Photobacterium phosphoreum* toxicity data index. *Water Pollut. Res. J. Can.* **26** (1991), 361-431.

[8] D. Van der Lelie, L. Regniers, B. Borremans, A. Provoost, L. Verschaeve, The VITOTOX® test, an SOS bioluminescence ***Salmonella typhimurium*** test to measure genotoxicity kinetics. *Mutat. Res.* **389** (1997), 279-290.

[9] P.F. Riska, Y. Su, S. Bardarov, L. Freulich, G. Sarkis, G. Hatful, C. Carriére, V. Kumar, J. Chan, W.R. Jr. Jacobs, Rapid film-based determination of antibiotic susceptibilities of Mycobacterium tuberculosis strains by using a luciferase reporter phage and the Bronx Box.. *J. Clin. Microbiol.,* **37** (1999), 1144-1449.

[10] J.B. Delehanty, K.M. Shaffer, B. Lin, B. A comparison of microscope slide substrates for use in transfected cell microarrays.*Biosens. Bioelectron.* **20** (2004), 773-779.

[11] J. R.Falsey, M. Renil, S. Park, S.J. Li, K.S. Lam, Peptide and small molecule microarray for high throughput cell adhesion and functional assays. *Bioconj. Chem.* **12** (2001), 346-53.

[12] J. Ziauddin, D.M. Sabatini, Microarrays of cells expressing defined cDNAs. *Nature* **411** (2001), 107-10.

[13] J.S.Andrews, V.P. Mason, I.P. Thompson, G.H. Markx, Investigation of the distribution of and interactions between microbial cells in artificially constructed biofilms by image analysis. 8th World Biosensor Congress; Granada, 2004.

[14] T.H. Rider, M.S. Petrovick, F.E. Nargi, J.D. Harper, E.D. Schwoebel, R.H. Mathews, D.J. Blanchard, L.T. Bortolin, A.M. Young, J. Chen, M.A. Hollis, A B cell-based sensor for rapid identification of pathogens. *Science* **301** (2003), 213-215.

[15] G. Moschopoulou, S. Kintzios, Application of "membrane-engineering" to bioelectric recognition cell sensors for the detection of picomole concentrations of superoxide radical: a novel biosensor principle.*Anal. Chim. Acta* **573-574** (2006), 90.

[16] S. Kintzios, E. Pistola, P. Panagiotopoulos, M, Bomsel, N, Alexandropoulos, F, Bem, C. Varveri, G. Ekonomou, J. Biselis, R. Levin, Bioelectric recognition assay (BERA). *Biosens. Bioelectron.* **16** (2001), 325-336.

[17] S. Kintzios, E. Pistola, J. Konstas, T. Matakiadis, N. Alexandropoulos, F. Bem, J. Biselis, R. Levin, Application of the Bioelectric recognition assay (BERA) for the detection of human and plant viruses: definition of operational parameters. *Biosens. Bioelectron.***16** (2001), 467-480.

[18] S. Kintzios, F. Bem, O. Mangana, K. Nomikou, P. Markoulatos, N. Alexandropoulos, C. Fasseas, V. Arakelyan, A-L. Petrou, K. Soukouli, G. Moschopoulou, C. Yialouris, A. Simonian, Study on the mechanism of Bioelectric Recognition Assay: evidence for immobilized cell membrane interactions with viral fragments. Biosens. Bioelectron. 20(2004), 906-916.

[19] S. Kintzios, E. Pistola, T. Matakiadis, C.P. Gialouris, N. Alexandropoulos, O. Maggana, Bioelectric recognition assay (BERA): a universal biosensory system for detecting viruses and evaluating antiviral agents.7th World Biosensors Conference, Kyoto, 2002.

[20] S. Kintzios, Ef. Makrygianni, E. Pistola, P. Panagiotopoulos, G. Economou, Effect of amino acids and amino acid analogues on the in vitro expression of glyphosate tolerance in Johnson grass (*Sorghum halepense* L. pers.) *J. Food, Agriculture and Environment* **3**(2003), 180-184.

[21] A. Zibert, E. Schreier, M. Roggendorf, Antibodies in human sera specific to hypervariable region 1 of hepatitis C virus can block viral attachment. *Virology* **208**(1995), 653-661.

[22] J.M.Romeo, , P. Ulrich, M.P.Busch, G.N. Vyas, Analysis of hepatitis C virus RNA prevalence and surrogate markers of infection among seropositive voluntary blood donors. *Hepatology* **17**(1993), 188-195.

[23] S. Kintzios, I. Marinopoulou, G. Moschopoulou, O. Mangana, K. Nomikou, K. Endo, I. Papanastasiou, A. Simonian, Construction of a novel, multi-analyte biosensor system for assaying cell division. *Biosens. Bioelectron.* **21** (2006), 1365-1373.

[24] D. Frossyniotis, Y. Anthopoulos, S. Kintzios, A. Perdikaris, C.P. Yialouris, A multisensor fusion system for the detection of plant viruses by combining artificial neural networks. *Lectures Comp. Sci.* **4132** (2006), 401-409.

[25] J. Slavik, *Fluorescent probes in cellular and molecular biology*, CRC Press: Boca Raton, 1993.

[26] V.A. Peechatinikov, F.F. Rizvanov, V.V. Pletnev, Permeability of sarcoplasma reticulum membrane for monovalent cations *Biofizika* **28** (1983), 669-673.

[27] A. Eddy, Use of carbocyanine dyes to assay membrane potential of mouse ascites tumor cells. Methods in Enzymol., 172 (1989), 95.

[28] R. Dixit, R. Cyr, Cell damage and reactive oxygen species production induced by fluorescence microscopy: effect on mitosis and guidelines for non-invasive fluorescence microscopy. Plant J. 36 (2003), 280-290.

[29] M.T. Osterholm, Getting prepared. *Foreign Affairs* **84** (2005), 24-37.

L.M. Lechuga et al. (Eds.)
IOS Press, 2008

Laser-Based Point Detector for on-line Identification of Biological Warfare Materials

Irina MOSKALENKO[1], Nikolay MOLODTSOV
Russian Research Center "Kurchatov Institute"
Moscow, Russia

Abstract. This paper reports the results of development and test of Laser-Based Point Detector (LBPD) biodetector for on-line identification of biological warfare materials. The test results of data evaluation of its potential to identify four microorganisms (spores of Bacillus anthracis and Bacillus cereus, vegetative cells of Brucella abortus and Escherichia coli) in pure and mixed cultures at varying concentrations on-line are presented. The LBPD was tested using wireless communications against not only simulants but pathogens as well. Identification specific pathogenic organisms on-line, in a semi-automated process without reagents achieve promising results (87%).

Keywords: Excimer laser, Spectra Photoluminescence, Biological materials, Multichannel Optical Detector, Recognition (MOD), Identification

[1]Corresponding author: Dr. Irina Moskalenko, Russian Research Center "Kurchatov Institute" *Moscow, Russia*. Tel.: +7095 196-7617, E-mail: ivm@nfi.kiae.ru

Introduction

It is well known that there are a lot of problems connected with environment contamination and decontamination. One of problems of special importance is detection and identification of contamination in real time just on-line and determination the board of contaminant area.

Detection of toxic species such as biological materials is a problem of special importance. Pollution of environment by these contaminants may be a result of technological accidents or terrorist attacks. Biological weapon agents (bacteria, viruses, toxins and genetically engineered species) are the most dangerous agents. It is necessary to take into consideration a possibility of water, food supplies, buildings, vehicles, human skin and etc. contamination. Serious problem exists to develop systems to detect toxic materials on-line for warning and protection of population against technological accidents or terrorist acts with biological agents. The main objective of these activities was to develop and test a detector based on application of excimer lasers for identification of toxic biological materials. The ability to accurately predict the dispersion, concentration, and ultimate fate of biological threat agents (BTAs) released into the environment in real time is essential to prepare for and respond to a biological threat agent release. Therefore, a key element of biological defense strategy is identification of the full set of biological threat agents (bacteria, viruses, fungi, and toxins) in the environment on-line [1-4]. However, development of biodetectors for field application is an extremely complicated task because of the variety of pathogenic microorganisms that could be used in an attack. Another problem is connected with a fact that many pathogenic organisms differ little from normal flora (sometimes only at the subspecies or strain level), a practical detection system has to discriminate between closely related organisms.

The effective testing of biological agents requires biosensor technology, which should be extremely sensitive, universal, reliable, and fast. Portability, rather than transportability, requires lighter weights, smaller volumes, moderate power consumption and enhanced energy efficiency [5-6]. Laser-based technologies can provide the basis for development of multiple-use devices, including point detectors, remote sensing (stand-off detectors), and non-invasive biosensors for point-of-care testing and diagnosis [7-9]. So it is importance to develop easily used methods and detectors easily operated, to be on duty and transported to measurement site very fast and begin to work without any alignment.

Approach and methodology

The technical approach and methodology of the LBPD is based on the use of laser induced photoluminescence of biological substances. Photoluminescence spectra excitation has been performed with pulse-periodic excimer laser for 3-D range resolved, providing a capability to carry out measurements of photoluminescence in wide spectral range including the near-UV and the visible sub ranges. The laboratory studies of spectral characteristics of biological materials by means of probing laser UV-radiation is an important part of total activity. In order to accelerate the work in this field, it is reasonable to use simulants. Such microorganisms have cell structures and chemical compositions being very close to similar parameters of bacteria- infecting agents of dangerous infections. The work with similar microorganisms does not present any danger and does not require special conditions to carry out experiments.

Description of the Laser-Based Point Detector

The portable Laser-Based Point Detector (**Figure 1**) includes an Excimer laser PSX-100 operating at 248 nm, a Multichannel optical detector (MOD), computer and software (Artificial Neural Network System). The multi-channel optical detector MOD-12 was designed for registration of spatial distribution of intensity of optical emission in one coordinate of its spectrum with its polychromator.

Three software programs were developed, evaluated and tested. The first program was used to collect spectral data from samples. This software was used both to develop a database of 3D spectra of standard concentrations of pure and mixture microorganisms, and to collect data from an unknown sample.

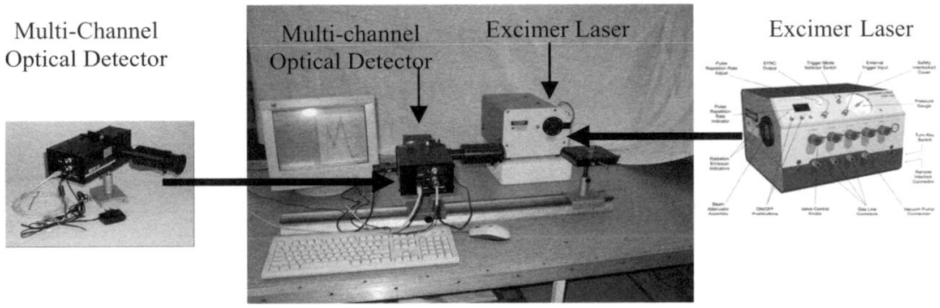

Figure. 1. Laser-Based Point Detector.

Excimer Laser PNX-100

Wavelength	248 nm
Pulse energy	5.5 mJ
Repetition rate	1-100 Hz
Pulse duration	2.5 ns
Beam dimensions	2x2 mm
Power requirements	220 V, 60 Hz
Gas Fill Lifetime	10^6 Shots
Dimensions (LxWxH), mm	300x260x210
Weight	13.3 kg
Flat-field spectrograph	
Spectral resolution	1 nm
High throughput; relative aperture	f/2.7
Dimensions (LxHxW), mm	110x115x75
Weight	1.2 kg

The data (spectrum of emitted light vs. time and the decay shape of the light emitted) was collected and visualized using the "MOD12" software, and stored in a ".LID" file (**Figure 2**). This file contained 20 time-slices of spectral data from a single shot. The second software was used to create a signature library of neural networks for recognizing the different microorganisms. Database includes: photoluminescence spectra for quartz cells, buffer media included (0,05M sodium phosphate buffer solution pH 7.4), distilled water, water pipe and water from puddle, amino acids and simulants of B.A., B.C., E.C., Br.Ab, and its mixture. The amount of measurement procedures includes more then 3,000 files (samples). The data was processed by a number of "neural networks" of data for each known type of bacterium and concentration. The Neural Network Software, incorporated a number of data files to create an analysis "library" for those files (**Figure 3**). The available "library" that could be searched consisted of 16 organisms x 4 concentrations. The third software used for bacteria recognition is shown in **Figure 4**.

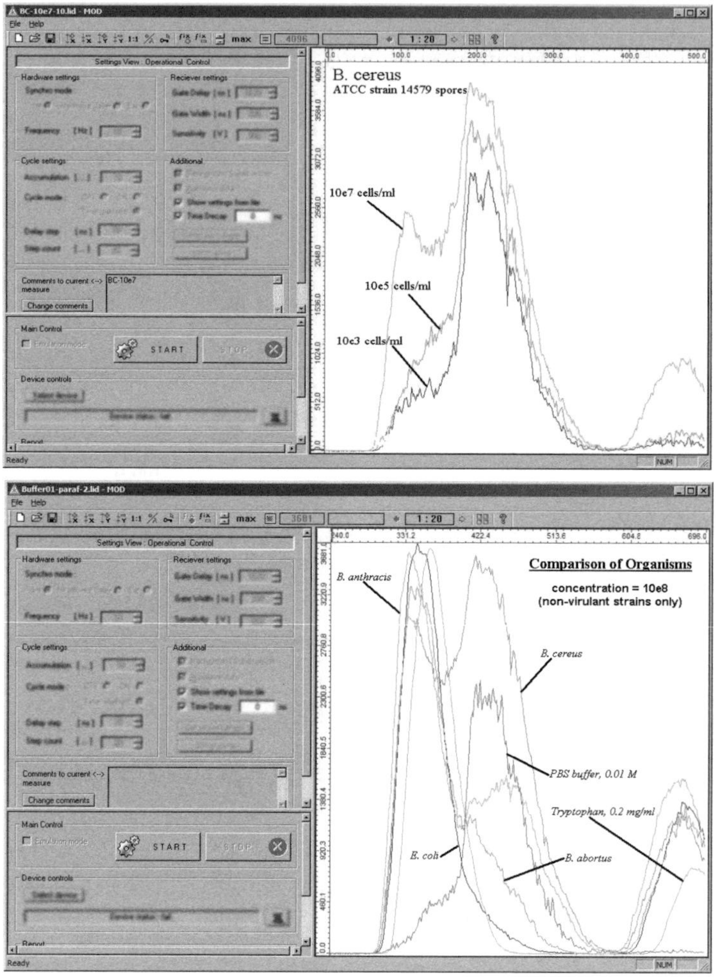

Figure. 2. Graphic user interface of the "MOD12" software.

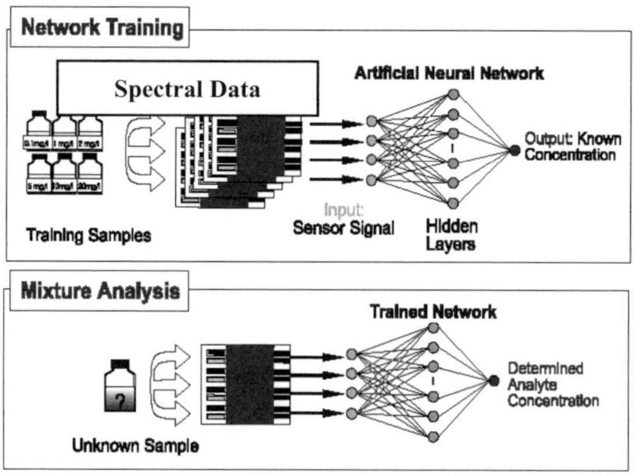

Figure. 3. The image of neural network software

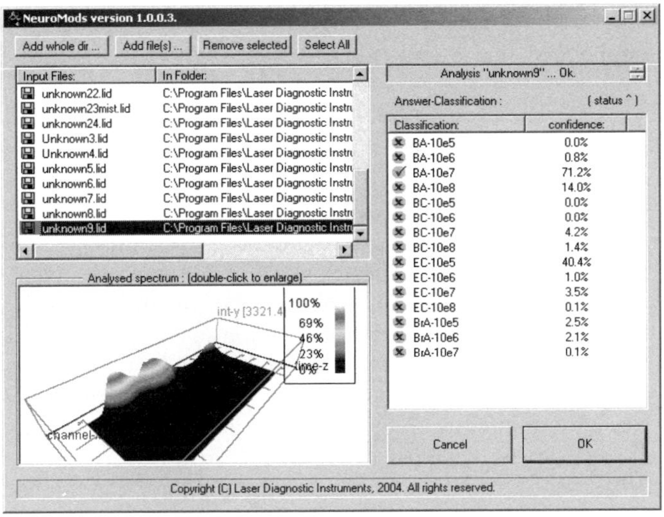

Figure. 4. Graphic user interface.

Biological agents

Spores
Bacillus anthracis Sterne vaccine strain, virulent strain - North Dakota

Bacillus cereus *American Type Culture Collection (ATCC) strain 14579*

Vegetative cells
Escherichia coli: ATCC strain 25922
Brucella abortus: Strain RB21– vaccine strain
Strain 2308 – virulent strain

Results and discussion

The results of testing demonstrated that the laser-based point detector potentially can be used for rapid detection and identification of different microorganisms based on ultraviolet (UV) laser-induced fluorescence (LIF) measurements with a satisfactory level of confidence. The "library" of signatures of different microorganisms was used to identify four pathogens in pure and mixed cultures. The spectral signatures contain not only information about the microorganism detected, but also information about the concentration of the microorganism. At this stage, a lower detection limit was 10^3cells/ml. The 3D fluorescence spectra illustrated the capability of the laser-based point detector to distinguish between different microorganisms. Receiving recognition for amino acids - 100% for simulants - about 87% using for training software. The detector operated very rapidly, with the time interval between sample interrogation (activating the laser) and sample analysis and recognition being a matter of a few seconds. The device used no special reagents, and power demands were moderate. The laser can be used for long periods, requiring refilling only after a few thousand shots, the laser gas required no special handling, and was commercially available from multiple gas suppliers.

Conclusions

The introduction of a third parameter, decay time, was used to produce 3-D Ultraviolet-LIF spectra, which appear to be quite promising for "fingerprinting" and discrimination of individual biological threat agent species alone and in mixtures using a novel LBPD prototype. The laser based technique has the advantages of simplicity, low cost, and rapidity and is particularly attractive for application to direct (label-free) detection of bacteria. The LBPD has good potential to distinguish between a bacterium under investigation and similar bacteria. The recognition software program at this stage of development was able to interpret 3-D spectra, which were readily demonstrated, with sufficient confidence levels, in most cases, to specifically identify and, discriminate among a small bacterial population under pure and mixture conditions. The operation of the wireless connection between computers in the Bio Safety Laboratory (BSL-3 and BSL-2 laboratories) was successful.

A lightweight laser-based biological agent point detector can be used for laboratory investigations (**Figure 5**), stationary (**Figure 6**) and hand point detector for mobile experts (e.g., for vehicles, buildings, soil, and human skin).

Operational capability of the Biodetector includes a stand-alone units directly at the point of sampling in remote environments for sudden changes in the concentration of suspicious bacteria in water, automatic transmission of biological alarm and data.

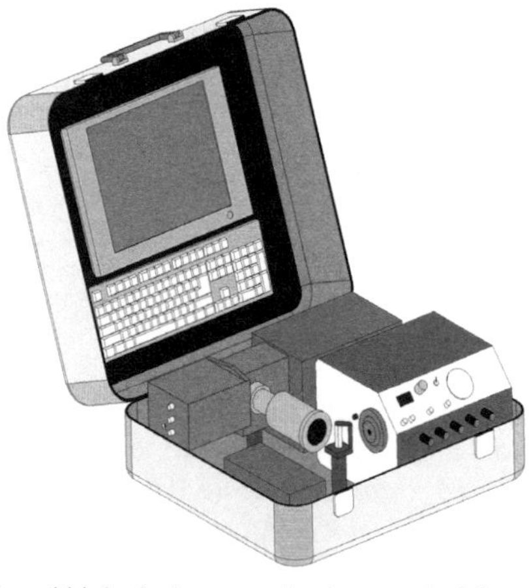

Figure. 5. A lightweight laser-based biological agent point detector for laboratory investigations.

Acknowledgements

This work was sponsored under a subcontract to the New Mexico Institute of Mining and Technology and Oklahoma State University through the U.S. Air Force Research Laboratories Air Expeditionary Forces Technologies Division and General Atomics Systems Engineering Division, under contract USAFRL Contract F08630-01-C-0065.

The work at the Kurchatov Institute was sponsored under a subcontract to the International Science and Technology Center (ISTC), Moscow, Russian Federation.

Figure. 6. Stationary point detector

References

[1] R.A. Greenfield, B.D. Lutz, M.M. Huycke, M.S. Gilmore, Unconventional biological threats and the molecular biological response to biological threats, Am. *J. Med. Sciences* **323** (2002), 350–357.

[2] N.J. Beeching, D.A.B. Dance, A.R.O. Miller, R.C. Spencer, Biological warfare and bioterrorism, *Brit. Med. J.* **324** (2002), 336–339.

[3] A.K. Deisingh, M. Thompson. Detection of infectious and toxigenic bacteria, *Analyst* **127** (2002), 567-581.

[4] D. Walt, D.R. Franz, Biological warfare detection, *Anal Chem.* **72** (2000), 739A–746A.

[5] S.S. Iqbal, M.W. Mayo, J.G. Bruno, B.V. Bronk, C.A. Batt, J.P. Chambers, A review of molecular recognition technologies for detection of biological threat agents, *Biosens. Bioelectron.* **15** (2000), 549–578.

[6] D. Ivnitski, D. O'Neil, A. Gattuso, R. Schlicht, M. Moore, M. Calidonna, R. Fisher, Nucleic acid approaches for detection and identification of biological warfare and infectious disease agents, *BioTechniques* **35** (2003), 862–869.

[7] E.T. Lagally, J.R. Scherer, R.G. Blazej, N.M. Toriello, B.A. Diep, M. Ramchandani, G.F. Sensabaugh, L.W. Riley, R.A. Mathies, Integrated portable genetic analysis microsystem for pathogen/infectious disease detection, *Anal. Chem.* **76** (2004), 3162–3170.

[8] A.M. Lefcourt, M.S. Kim, Y.R. Chen, Automated detection of fecal contamination of apples by multispectral laser-induced fluorescence imaging, *Appl. Optics* **42** (2003), 3935–3943.

[9] P. Francois, M. Bento, P. Vaudaux, J. Schrenzel, Comparison of fluorescence and resonance light scattering for highly sensitive microarray detection of bacterial pathogens, J. Microbiol. Methods 55 (2003), 755–762.

L.M. Lechuga et al. (Eds.)
IOS Press, 2008

Pre-symptomatic Prediction of Illness in Mice Inoculated with Cowpox

James.R. KERCHER[a], Bill W. COLSTON[a], Jr., Richard G. LANGLOIS[a], C. Rick LYONS[b] and Fred P. MILANOVICH[a]
[a]*Lawrence Livermore National Laboratory, LIV, CA, USA*
[b]*University of New Mexico, ABQ, USA*

Abstract. We describe here research directed towards early (presyndromic) diagnosis of infection. By using a mouse model and a multi-component blood protein diagnostic tool we detected cowpox infection several days in advance of overt symptoms with high accuracy. We provide details of the experimental design and measurement technique and elaborate on the long-range implication of these results.

Keywords. Biodefense, disease, diagnostic, mouse model, pox

Introduction

There is growing urgency to develop techniques for rapid detection and diagnosis of infectious disease in human populations. Rapid detection is critical for reducing the morbidity and mortality from either bio-terrorism events or newly emerging diseases. Current methods for direct agent detection using culture methods or microbial component detection using antibodies or PCR have a number of limitations. Rapid microbial detection in blood may not be possible for agents that remain localized to the site of infection or agents that do not appear in peripheral blood until the later stages of the disease. Cell culture and sample enrichment procedures can also require several days. Finally, newly emerging diseases or genetically modified organisms may have never been seen before complicating organism-specific detection methods.

Host responses may provide early signals in blood even from localized infections. Cells in the innate immune system produce a rapid response after initial contact with a potential pathogen. While pathogen responses initially involve local cell signaling processes designed to activate near-by immune cells, cascades of cytokines and chemokines are released into the periphery to activate additional cells types and to cause them to migrate to the site of the infection. Thus, early immune responses may provide general indicators for the presence of many different infection types. A spectrum of innate and adaptive immune markers, in combination with other biochemical markers comprise a 'signature' that may allow for timely, disease-specific detection.

Direct studies of the time course of natural diseases in humans is complicated by the difficulty in defining exposure doses and exposure timing. Model systems, both in cell culture and animal models, allow precise control over exposure dose and timing. Studies of specific tissue types in culture have been used to define early responses in cells to pathogen exposure, but it is often difficult to relate these tissue-specific responses to systemic responses in the whole animal. Consequently, we chose cowpox infection in mice as the model system for studies of early disease detection through blood protein signatures. Mice have been shown to be susceptible to infection by both cowpox and vaccinia viruses [1, 2]. The severity of infection varies from mild to lethal depending on the strain of mouse, strain of virus, and the location and dose of viral challenge. Viral instillation in the lung of mice produces a pulmonary infection that has been used as a model for pulmonary smallpox infection in humans, so that this model has been used extensively for studies of anti-viral drugs [3, 4].

The TK- strain of cowpox virus into BALB/c mice model used for these studies exhibits three major features important to early detection of infections. First, there is an incubation period of about 6 days before the mice show signs of illness. This provides the opportunity to assess protein changes in serum throughout this prodromal period, as well as the period of active infection. Second, this model produces a localized infection in the lung, with no live virus detectable in the blood using plate assays. Thus, this model is well suited to analyze whole-animal systemic responses in blood to a localized infection. Finally, the conditions used for these studies produce a serious illness, but no lethality was observed from viral infection. This feature allows assessment of the early stages of recovery from infection and insures that the biochemical changes we observe reflect moderate disease rather than the severe toxicity of super-lethal doses.

In this paper we describe the use of canonical variate analysis (CVA) to predict pre-symptomatic illness in mice inoculated with cowpox virus. Signatures were developed from measurements of specific blood serum protein concentrations determined by Rules Based Medicine, Inc., Austin, TX, using Luminex Liquid Array LA technology. We applied CVA to training sets of data to find transformations to the canonical space specifying the optimal group separation. These transformations were then applied to test sets to determine where individual mice mapped to in canonical space. Group membership of each individual was predicted by assigning membership to the training group lying closest to the individual in canonical space. Finally, assigned membership was compared with known true membership and the results expressed as a confusion matrix.

We have made three separate analyses and have recommendations for follow on analyses based on the results for the two experiments analyzed here.

Experiments

All experimental work with the mice was performed in a biosafety level 3 (BSL-3) animal facility at the University of New Mexico [5, 6]. Mice were infected with CPV by surgical intratracheal instillation. The infected mouse group received 50 ul of media (PBS 2.5% BGS diluted 1:2 in Tris buffer) containing CPV. The sham group received the same intratracheal surgical procedure with 50ul of media only. The naïve control group received no surgery or CPV.

Animals were sacrificed at each experimental time point to obtain serum samples. The animals were euthanized via Avertin overdose. Blood was collected from clipped vena cava and placed in eppendorf tube. The blood was allowed to coagulate for at least 30 minutes at 4°C. The whole blood was then centrifuged at 4500 RPM for 10 min in microfuge, and the serum from each mouse was placed in individual eppendorf tubes. The samples were stored at -80°C and analyzed at the end of the study. Serum volumes varied from 700 to 200 microliters, with lower recoveries at the peak of the disease.

The two experiments analyzed herein were conducted a month apart in spring and summer of 2005. In each experiment 60 mice were given tracheotomies. In 30 of these mice, cowpox virus (CPV) was injected in the tracheotomies. In the other 30 mice, saline solutions were injected (sham). On days 2, 3, 4, 5, 6, and 8, five CPV and five sham mice were sacrificed and blood collected from the chest cavities. Mouse serum samples were transported frozen on dry ice to Rules Based Medicine (RBM) for multiplex immunoassay analysis (Rules Based Medicine, Inc., Austin, TX). RBM utilizes a multi-analyte panel to quantify the concentrations of about 60 host antigens in mouse serum samples. A total of 50 microliters of serum were shipped for each analysis, with all serum samples coded for blind analysis. While some aspects of this multiplex analysis are proprietary, the following briefly describes the approach. Sample is incubated with a mixture of fluorescently labeled microsphere types, with each bead type conjugated to a different capture antibody. A mixture of biotinylated secondary antibodies is then added to label bead-captured antigen. Finally streptavidin-phycoerythrin is added to fluorescently label the captured antigen. Flow cytometric analysis with a Luminex 100 flow analyzer is used to quantify fluorescence signals for ach antigen in the analysis. Purified antigen standards are included in some samples to develop standard curves for relating bead-based fluorescence to antigen concentration. Sample processing typically requires 1 to 3 hours, and flow cytometric analysis takes about 1 minute per sample [7, 8, 9]. Additional details of the RBM analysis can be found at www.rulesbasedmedicine.com .

Assay results are reported in units of protein concentration. A "Lowest Detectable Dose" (LDD) is also reported for each antigen. The LDD is the antigen concentration that produces a fluorescence signal that is 3 standard deviations above the fluorescence of negative control beads. Table 1 below lists the 54 proteins for which we had assays, of these we used 43 proteins in the analysis described herein. These 43 proteins are the ones for which at least two mice had levels above the LDD in both experiments. Protein concentration values were set to 0.5*LLD for assay results that were below the LDD for the protein.

Animals were monitored for behavioral signs of illness including ruffled fur and decreased activity. While a few animals began to show symptoms on day 5, most animals showed clear signs of illness on days 6 and 8.

Table 1. Proteins measured in the RBM panel used for experiments 1 and 2.

Apo A1 (Apolipoprotein A1)	IL-1alpha (Interleukin-1alpha)	MDC (Macrophage-Derived Chemokine)
CRP (C Reactive Protein)	IL-1beta (Interleukin-1beta)	MIP-1alpha (Macrophage Inflammatory Protein-1alpha)
EGF (Epidermal Growth Factor)	IL-2 (Interleukin-2)	MIP-1beta (Macrophage Inflammatory Protein-1beta)
Endothelin-1	IL-3 (Interleukin-3)	MIP-1gamma (Macrophage Inflammatory Protein-1gamma)
Eotaxin	IL-4 (Interleukin-4)	MIP-2 (Macrophage Inflammatory Protein-2)
Factor VII	IL-5 (Interleukin-5)	MIP-3beta (Macrophage Inflammatory Protein-3beta)

FGF-9 (Fibroblast Growth Factor-9)	IL-6 (Interleukin-6)	Myoglobin
FGF-basic (Fibroblast Growth Factor-basic)	IL-7 (Interleukin-7)	OSM (Oncostatin M)
GCP-2 (Granulocyte Chemotactic Protein-2)	Insulin	RANTES (Regulation Upon Activation, Normal T-Cell Expressed and Secreted)
GM-CSF (Granulocyte Macrophage-Colony Stimulating Factor)	IP-10 (Inducible Protein-10)	SCF (Stem Cell Factor)
Haptoglobin	KC/GROalpha (Melanoma Growth Stimulatory Activity Protein)	SGOT (Serum Glutamic-Oxaloacetic Transaminase)
IFN-gamma (Interferon-gamma)	Leptin	TIMP-1 (Tissue Inhibitor of Metalloproteinase Type-1)
IgA (Immunoglobulin A)	LIF (Leukemia Inhibitory Factor)	Tissue Factor
IL-10 (Interleukin-10)	Lymphotactin	TNF-alpha (Tumor Necrosis Factor-alpha)
IL-11 (Interleukin-11)	MCP-1 (Monocyte Chemoattractant Protein-1)	TPO (Thrombopoietin)
IL-12p70 (Interleukin-12p70)	MCP-3 (Monocyte Chemoattractant Protein-3)	VCAM-1 (Vascular Cell Adhesion Molecule-1)
IL-17 (Interleukin-17)	MCP-5 (Monocyte Chemoattractant Protein-5)	VEGF (Vascular Endothelial Cell Growth Factor)
IL-18 (Interleukin-18)	M-CSF (Macrophage-Colony Stimulating Factor)	vWF (von Willebrand Factor)

Data Analysis

To analyze the training sets we used standard CVA as discussed in Krzanowski and Srivatava [10, 11]. These analyses were done in a backward elimination selection method, which proceeds using the test on the change in Wilks ratio for each elimination as proposed by Rao [12] and described in detail by Hawkins and McHenry [13, 14]. See detail on signature development and class prediction in Appendix I.

The data was analyzed by considering the CPV mice sacrificed on days 2 and 3 as one group, the CPV mice for days 4 and 5 as another group, the CPV mice for days 6 and 8 as a third group, the sham mice for days 2 and 3 as the fourth group, the sham mice for days 4 and 5 as the fifth group, and the sham mice for days 6 and 8 as the sixth group. After the training set is analyzed to find the positions of the training mice (and groups) in canonical space, the "unknown" individuals from training set are then assigned to one of these six groups by mapping them to canonical space and finding the nearest training group for each test individual.

We have made three separate analyses. First we used experiment 1 (the first experiment performed) as a training set and experiment 2 as the test set. Second we used experiment 2 as the training set and experiment 1 as the test set. Third we made a

10-fold external cross validation on both experiments. In this cross validation method, we combine both experiments into one data set. Thus there will be the same six groups but each group will have twenty members instead of the ten when the experiments are considered separately. Then the combined data set is repeatedly reanalyzed by random partitioning followed by training and testing. This method provides an estimate of the overall error rate in predicting class membership. In our case assigning class membership is equivalent to deciding whether or not an individual mouse is infected with CPV and how much time has elapsed since the infection (or tracheotomy in the case of sham mice).

Details on the class prediction and cross validation techniques are given in Appendix I and II.

Results

Training Set: = Experiment 1 / Test Set = Experiment 2

In this analysis we found that eight proteins significantly separated the six groups in the training set. The proteins were:

Proteins used in CVA of experiment 1	
MCP-5_(Monocyte_Chemoattr)	KC_GROalpha_(Melanoma_Gro)
Haptoglobin	IL-1alpha_(Interleukin-1a)
MIP-1beta_(Macrophage_Inf)	Myoglobin
IL-18_(Interleukin-18)	MMP-9_(Matrix_Metalloprot)

The canonical transformations were constructed from the concentrations of these eight proteins.

In Table 2 we show the confusion matrix for using experiment 1 as the training set and using experiment 2 as the test set. To read the table, read each row from left to right. We see immediately that while pox mice for days 4-5 and days 6-8 were well predicted to be the appropriate pox group, sham mice were overwhelmingly predicted to be pox mice. This level of false positives renders this particular transformation useless. The variables (proteins) that separated the groups in experiment 1 were peculiar to experiment 1, possibly due to some irreproducible idiosyncrasy of the conditions under which experiment 1 was conducted.

Table 2. Confusion matrix for training CVA using experiment 1. Group assignments made on experiment 2. Column labeled by "Total" contains the number of mice in groups labeled in column one labeled by "Group"

Group	**Total**	**Sham 2-3 days**	**Sham 4-5 days**	**Sham 6-8 days**	**Pox 2-3 days**	**Pox 4-5 days**	**Pox 6-8 days**
Sham 2-3 days	10	0	0	0	4	6	0
Sham 4-5 days	10	0	0	0	1	9	0
Sham 6-8 days	10	0	0	3	1	5	1

Pox 2-3 days	10	0	0	0	4	6	0
Pox 4-5 days	10	0	0	0	0	9	1
Pox 6-8 days	10	0	0	0	0	1	9

Training Set = Experiment 2 / Test Set = Experiment 1

In this analysis we found that eight proteins significantly separated the six groups in the training set. The proteins were:

Proteins used in CVA of experiment 2	
MIP-1beta_(Macrophage_Inf	Haptoglobin
IP-10_(Inducible_Protein-	GCP-2_(Granulocyte_Chemot
CD40_Ligand	Leptin
MCP-5_(Monocyte_Chemoattr	Growth_Hormone

The canonical transformations were constructed from the concentrations of these eight proteins. Note that there is some overlap with the proteins chosen in the first training analysis. There is also some discrepancy between the two sets.

In Table 3 we show the confusion matrix for using experiment 2 as the training set and using experiment 1 as the test set. To read the table, read each row from left to right. We see immediately that sham mice are overwhelmingly mapped to sham mice. We see that pox mice in the 4-5 day group are mainly mapped to the correct pox group and pox mice in the 6-8 day groups are mainly mapped to pox mice. We did not do well predicting the exact group for 2-3 day pox mice.

Table 3. Confusion matrix for training CVA using experiment 2. Group assignments made on experiment 1. Column labeled by "Total" contains the number of mice in groups labeled in column one labeled by "Group".

Group	**Total**	**Sham 2-3 days**	**Sham 4-5 days**	**Sham 6-8 days**	**Pox 2-3 days**	**Pox 4-5 days**	**Pox 6-8 days**
Sham 2-3 days	10	5	3	0	2	0	0
Sham 4-5 days	10	2	1	3	4	0	0
Sham 6-8 days	10	2	5	3	0	0	0
Pox 2-3 days	10	2	1	0	7	0	0
Pox 4-5 days	10	0	0	1	1	8	0
Pox 6-8 days	10	0	0	0	2	0	8

We regard the above results as extremely promising as most pox mice in the 4-5 day group are correctly mapped, even though most animals show no signs of illness at this time period.

Cross Validation

In the 10-fold external cross validation we found that seventeen proteins were selected as significantly separating the six groups.

In Table 4 we show the confusion matrix for the cross validation. Shams are mapped mainly to shams. We see that pox mice in the 4-5 day group and in the 6-8 day group are mainly mapped to the correct group of pox mice. We did not do well predicting the exact group for 2-3 day pox mice. Many 2-3 day pox mice are mapped to 2-3 day sham mice. While most pre-symptomatic animals at 4-5 days can be correctly classified using these 17 proteins, protein alterations at 2-3 days may be too small to differentiate pox from sham animals.

These results show promise for disease detection when one considers that which sham-group a sham mouse belongs to is of little concern as long as we know it is a sham mouse. However which group a pox mouse belongs to is of concern because of treatment options (in other diseases) or public health considerations.

Table 4. Cross validation confusion matrix

Group	Total	Sham 2-3 days	Sham 4-5 days	Sham 6-8 days	Pox 2-3 days	Pox 4-5 days	Pox 6-8 days
Sham 2-3 days	20000	12639	2087	1398	3245	631	0
Sham 4-5 days	20000	2974	9658	6534	833	1	0
Sham 6-8 days	20000	332	8161	10185	771	0	551
Pox 2-3 days	20000	6462	155	93	13289	0	1
Pox 4-5 days	20000	161	123	124	1762	17597	233
Pox 6-8 days	20000	17	1	0	1746	144	18092

Conclusion

In spite of the aggressively invasive nature of the inoculation, we were able to see the pattern of expressed proteins relax to normal then signal an alert to infection prior to symptoms. This gives promise for the development of a cost effective triage diagnostic that would optimize response to a disease epidemic or pandemic. The next step in the research is to compare a second disease to the cowpox protein signature and confirm that selectivity is possible.

Acknowledgement

This work was performed under the auspices of the U.S. Department of Energy by University of California, Lawrence Livermore National Laboratory under Contract W-7405-Eng-48. Research funding was provided by the LLNL Laboratory Directed Research and Development Program. We thank Dr. Jim Mapes of Rules Based Medicine, Austin, TX for helpful discussions. We also acknowledge the helpful discussions with Carl Melius and Stephen Johnston.

References

[1] M.J. Martinez, M.P. Bray, and J.W. Huggins, A mouse model of aerosol-transmitted orthopoxviral disease, *Arch Path Lab Med* **124** (2000) 362–377.
[2] P.C. Reading and G.L. Smith, A kinetic analysis of immune mediators in the lungs of mice infected with vaccinia virus and comparison with intradermal infection, *J Gen Virol* **84** (2003) 1973–1983.
[3] D.C. Quenelle, D.J. Collins, W.B. Wan, J.R. Beadle, K.Y. Hostetler, and E.R. Kern, Oral treatment of cowpox and vaccinia virus infections in mice with ether lipid esters of cidofovir, *Antimicrob Agents Chemother*, **48(2)** (Feb. 2004) 404–12. Erratum in: *Antimicrob Agents Chemother*, **48(5)** (May 2004) 1919.
[4] D.C. Quenelle, D.J. Collins, and E.R. Kern, Efficacy of multiple- or single-dose cidofovir against vaccinia and cowpox virus infections in mice, *Antimicrob Agents Chemother*, **47(10)** (Oct. 2003) 3275–3280.
[5] C.R. Lyons, J. Lovchik, J. Hutt, M.F. Lipscomb, E. Wang, S. Heninger, L. Berliba, and K. Garrison, Murine model of pulmonary anthrax: kinetics of dissemination, histopathology, and mouse strain susceptibility, *Infect Immun.*, **72(8)** (Aug. 2004) 4801–4809.
[6] A.M. Talaat, R. Lyons, S.T. Howard, and S.A. Johnston, The temporal expression profile of mycobacterium tuberculosis infection in mice, *Proc Natl Acad Sci USA*, **101** (2004) 4602–4607
[7] M.T. McBride, D. Masquelier, B.J. Hindson, A.J. Makarewicz, S. Brown, K. Burris, T. Metz, R.G. Langlois, K.W. Tsang, R. Bryan, D.A. Anderson, K.S. Venkateswaran, F.P. Milanovich, and B.W. Colston Jr., Autonomous detection of aerosolized Bacillus anthracis and Yersinia pestis, *Anal Chem.*, **75(20)** (Oct. 15, 2003) 5293–5299.
[8] M.T. McBride, S. Gammon, M. Pitesky, T.W. O'Brien, T. Smith, J. Aldrich, R.G. Langlois, B. Colston, and K.S. Venkateswaran, Multiplexed liquid arrays for simultaneous detection of simulants of biological warfare agents, *Anal Chem.* **75(8)** (Apr. 15, 2003) 1924–1930.
[9] U. Prabhakar, E. Eirikis, and H.M. Davis, Simultaneous quantification of proinflammatory cytokines in human plasma using the LabMAP assay,. *J Immunol Methods*, **260(1–2)** (Feb. 1, 2002) 207–218.
[10] W.J. Krzanowski, *Principles of multivariate analysis*, Clarendon Press, Oxford, 2000.
[11] M.S. Srivastava, *Methods of multivariate statistics*, Wiley, 2002.
[12] C.R. Rao, Inference on discriminant function coefficients, In *Essays on Probability and Statistics* (R.C. Bose, I.M. Chakravarti, P.C. Mahalanobis, C.R. Rao, and K.J.C. Smith, eds.) University of North Carolina Press, Chapel Hill, NC, 1970, 587–602.
[13] D.M. Hawkins, The subset problem in multivariate analysis of variance, *J. Royal Statist. Soc. B*, **38** (1976) 132–139.
[14] C.E. McHenry, Computation of a best subset in multivariate analysis, *Appl. Statist.*, **27** (1978) 291–296.
[15] C. Ambroise, G. McLachlan. 2002. Selection bias in gene extraction on the basis of microarray gene-expression data. Proc. National Academy Sci. 99:6562-6566.

Appendix I. Class Prediction by Canonical Analysis

Denote the number of groups as h, the number of observations as N, and the number of variables as p. Divide the data set into a training set and a test set with observation numbers N_r and N_s, respectively, and $N=N_r+N_s$. To perform class prediction on a test set using canonical analysis, first form the between-group sum-of-squares-and-cross-products matrix (SSCP), denoted by **B**, and the within-group SSCP **W** from the training-set data matrix. Then solve $\mathbf{Be} - \lambda\mathbf{We}=0$ for the eigenvalue λ and the eigenvector **e**. There will be min(p, h–1) nonzero eigenvalues. The associated eigenvectors are the only ones we are concerned with. These eigenvectors map both the training set and test set observations to canonical space, i.e., the position of an observation on the axis defined by the ith eigenvector is $y_i=\mathbf{e}^{i\mathbf{T}}\mathbf{x}_j$ where **T** denotes the transpose and $\mathbf{x}_j$ denotes the vector in RBM protein measurements of a particular observation j. All the xj's from a particular group in the training set are used to define the centroids (mean position of the kth group is μ_k) of the group in canonical space. Each $\mathbf{x}_j$ from the test set is assigned membership in the group whose centroid is nearest to it in canonical space, i.e., **y** belongs to the kth group if $||\mathbf{y} - \mu_k||$ is the minimum for $||\mathbf{y}-\mu_i||$ for all i.

In the backward elimination procedure, we begin with p proteins and eliminate the protein contributing the least to the discrimination. This continues successively until only one protein is left. At each stage we calculate the canonical analysis and assign membership of the test set observations. So at each stage we have the number of incorrect assignments for that number of proteins used in the canonical analysis.

Appendix II. 10-Fold External Cross Validation and Data Handling

Cross Validation

In an M-fold external cross validation, the data set is partitioned into M subsets of approximately equal size. Each subset is selected in turn as the test set, the remaining M–1 subsets are combined as the training set. Then the discrimination method is applied to the training set and any variable selection is carried out for that training set. After the discrimination function is constructed, we use the discrimination function to assign group membership to the observations in the test set. This procedure is repeated for all the M subsets. The distinction between an external cross validation and an internal cross validation is that in the former case variable selection occurs each time a discrimination function is constructed. In the latter case, variable selection is done only once at the outset and then the same variables are used for all construction of discrimination functions. Because the full data set is partitioned randomly into the M subsets, one must perform the portioning many times to get a sense of the statistical variation in the error rate of group assignment. In the cases studied here, we performed 1000 partitions. Ambroise and MacLachlan [15] have established the value of the 10-fold external cross validation to estimate of the error rate.

L.M. Lechuga et al. (Eds.)
IOS Press, 2008

PQQ-dehydrogenases as a Favorable Components for Biosensor Design

Valdas LAURINAVIČIUS[1], Rolandas MEŠKYS, Julija RAZUMIENĖ, Bogumila KURTINAITIENĖ
Institute of Biochemistry, Vilnius, Lithuania.

Abstract. PQQ-dependent glucose, alcohol and glycerol dehydrogenases were purified and used for the design of a number of biosensors. Platinum and carbon paste electrodes were used for the design of the biosensor design. A number or soluble and polymer-type mediators have been tested. Biosensors possessing a direct electron transport from the active center of enzymes to the surface of the carbon paste electrode were developed.

Keywords. PQQ-dehydrogenases, electrochemical biosensors

[1]**Corresponding Author**: Mokslininku 12, Vilnius, LT-08662, Lithuania,
E-mail: valdasL@bchi.lt

Introduction

Forty years ago a new class of dehydrogenases was reported [1]. However, only after twenty years the structure of a prosthetic group was elucidated and referred as pyrroloquinoline quinine (PQQ) [2]. The substrate reduces PQQ by two electrons through a radical semiquinone step at the potential (+90 mV) that is higher as compared with NAD^+ or FAD:

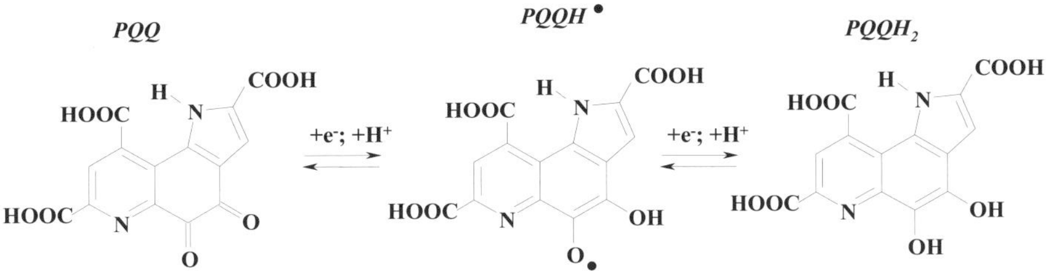

Natural electron acceptors for this type of enzymes are ubiquinones and cytochromes, but not oxygen [3-4]. A number of artificial mediators can regenerate the reduced active center of PQQ-dehydrogenases. Derivatives of organic metal complexes - ferrocenes [5], organic complexes of osmium and ruthenium [6,7], as well as a number or organic heterocyclic compounds [8], or insoluble redox polymers [9-10] are good mediators between the active center of enzyme and the electrode surface.

PQQ-dehydrogenases recognize aliphatic alcohols (alcohol dehydrogenase, ADH), glycerol (glycerol dehydrogenase, Glyc-DH), and sugars (glucose (aldose) dehydrogenase, GDH), some of them oxidize aliphatic amines. Some quinoenzymes are located in cytoplasma or periplasma as soluble globular enzymes (s-GDH or s-ADH). Other types of the PQQ-dehydrogenases are located in membrane (m-GDH, and m-ADH, Glyc-GH). They are different enzymes with different size, mechanism of action and substrate specificity.

A short review about purification and application of a number PQQ-GDH, PQQ-ADH and Glyc-DH was reported recently [11]. This communication summarized the results of direct electron transport in PQQ-dehydrogenases and an application of this phenomena in biosensor design.

Glucose Dehydrogenases in Biosensor Design

There are two known types of PQQ-glucose dehydrogenases produced by *Acinetobacter calcoaceticus* L.M.D. 79.41: soluble (s-GDH) and bound to the outer surface of the cytoplasmic membrane (m-GDH) [12]. sPQQ-GDH from *Acinetobacter calcoaceticus* has been cloned and a recombinant enzyme was expressed in *Escherichia coli* strain.

s-GDH from *Acinetobacter* calcoaceticus L.M.D. 79.41 was purified according to the known protocol [13]. Partially purified s-GDH had a specific activity of 250 U/mg protein. D-glucose (1 mM) was used as a default substrate. m-GDH was purified from *Erwinia* sp. 34-1 [14], (specific activity 12 U/mg). The enzyme was used as suspension in 0.02 mol/l potassium phosphate buffer (pH 7.0) containing 10% glycerol.

GDH–based biosensors were designed on the basis of carbon electrodes modified with ferrocene derivatives [5]. Recently special carbon black was synthesized at the Faculty of Chemistry of Vilnius University and used for the preparation of carbon paste electrochemical GDH-based biosensors possessing direct electron transport [15].

Electrochemical parameters (data on Figures 1 and 2) of both types of biosensors and an electrochemical behavior of PQQ showed that rate limiting step is a electrochemical reduction of the same group – PQQ.

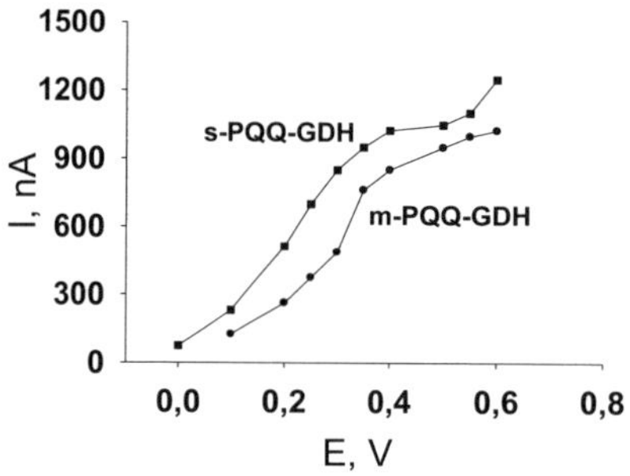

Figure 1. Voltammograms s-GDH and m-GDH adsorbed on the carbon electrode possessing a direct electron transport.

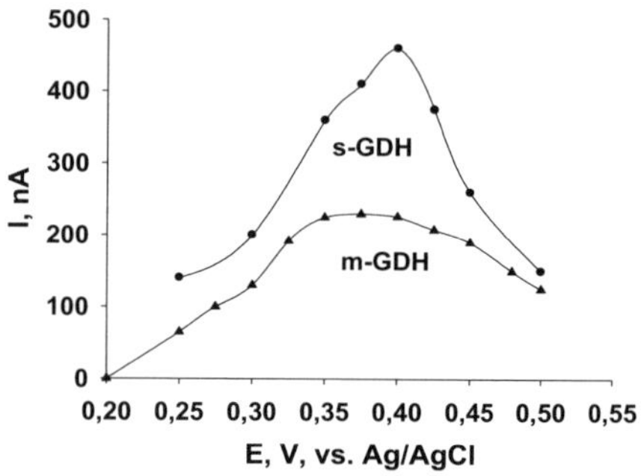

Figure 2. Voltammograms of s-GDH, and m-GDH adsorbed on the carbon electrode modified with 4- ferrocenylphenol.

However, the substrate specificity of the native and immobilized enzymes changes during immobilization in different ways. In the case of s-GDH an immobilization decreases the substrate selectivity of the enzyme, when immobilization of m-GDH improves the substrate specificity. Data are presented in Table 1.

Table 1. Substrate specificity (%) of s-GDH ir m-GDH

Substrate	s-GDH,	s-GDH-ferrocene,	s-GDH-carbon paste	m-GDH	m-GDH-ferrocene	m-GDH-carbon paste
Glucose	100	100	100	100	100	100
Galactose	38	98	97	76	7.4	8
Arabinose	37	88	80	28	4.4	6
Maltose	86	99	95	1.1	0.5	0.8
Cellobiose	69	67	69	0	0.1	0.1
Lactose	72	99	90	0	0	0

It is well known that PQQ dissociates easily from s-GDH active center [16]. This data was also confirmed when s-GDH was immobilized on a cystamine-modified gold electrode [17]. On the basis of these data we have suggested the following mechanism of action of immobilized s-GDH. PQQ dissociates from the active center of enzyme and acts as a free mediator between the enzyme active center and the electrode. Due to a high activity of the s-GDH, the rate-limiting step of the process becomes the binding of PQQ to the apo-enzyme. During the biosensor aging, the sensitivity of the biosensor decreases, probably due to the loose of the PQQ from the outer layers of the immobilized enzyme on the surface of the biosensor. Thus, the distance between electrode surface and active center of enzyme is reducing, and, in the case of s-GDH adsorbed on carbon paste electrode when direct electron transport is observed, $K_{M(app.)}$ of the immobilized enzyme is decreasing (Figure 3).

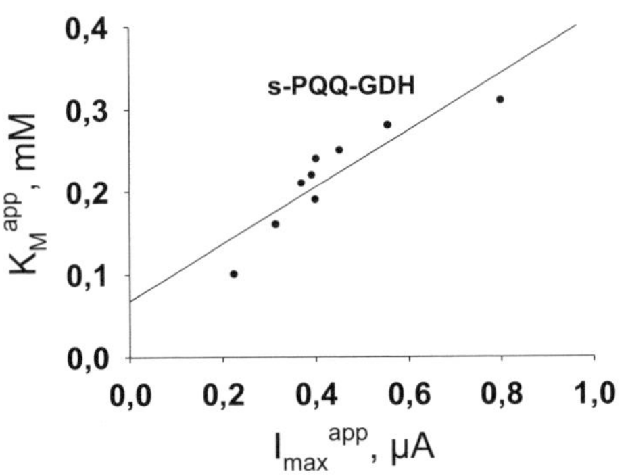

Figure 3. Dependence of $K_{M(app.)}$ of the s-GDH adsorbed on the carbon paste electrode on the decrease of biosensor sensitivity.

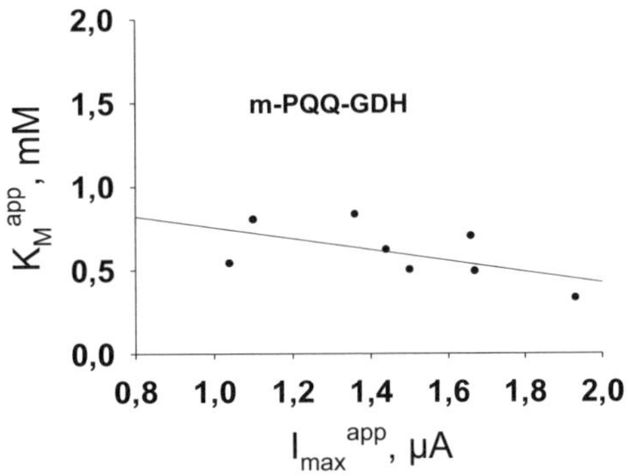

Figure 4 Dependence of $K_{M(app.)}$ of the m-GDH adsorbed on the carbon paste electrode on the decrease of biosensor sensitivity.

PQQ in the active center of m-GDH is bound more tightly than in s-GDH. During the immobilization the enzyme shows an improvement in substrate selectivity and a significant increase in K_M for glucose during the aging of the biosensor. In this case m-

GDH adsorbs on the carbon paste electrode in the same manner as on the hydrophobic membrane. The enzyme is anchored to the surface of the electrode *via* the hydrophobic anchoring subunit and the enzyme globule operates in some distance from the surface. Thereby, it is believable, that the immobilization of m-GDH on the carbon paste mimics the natural surrounding of the enzyme. It was shown that m-GDH forms a substrate-enzyme-acceptor complex [18]. The rate of the regeneration of the active center is probably faster as compared to that for the native enzyme in solution and the rate of the substrate binding starts to play a more significant role in the overall process, thereby improving a substrate selectivity. During the aging the conformational changes would deform the active center of the enzyme, thereby binding of the substrate might be affected, that would lead to the increase of $K_{M(app.)}$.

In many papers it was stated that biosensors on the basis of PQQ-dehydrogenases should be insensitive to the oxygen fluctuations in the sample. We carried out a model experiment using two glucose biosensors simultaneously. One sensor was glucose oxidase based and the second was based on s-GDH immobilized on the carbon paste. The sensors were inserted into a reactor containing growing yeasts *Saccharomyces cerevisiae*. An oxygen consumption was measured by an electrochemical oxygen sensor additionally installed into the reactor. Data are presented in Figure 5.

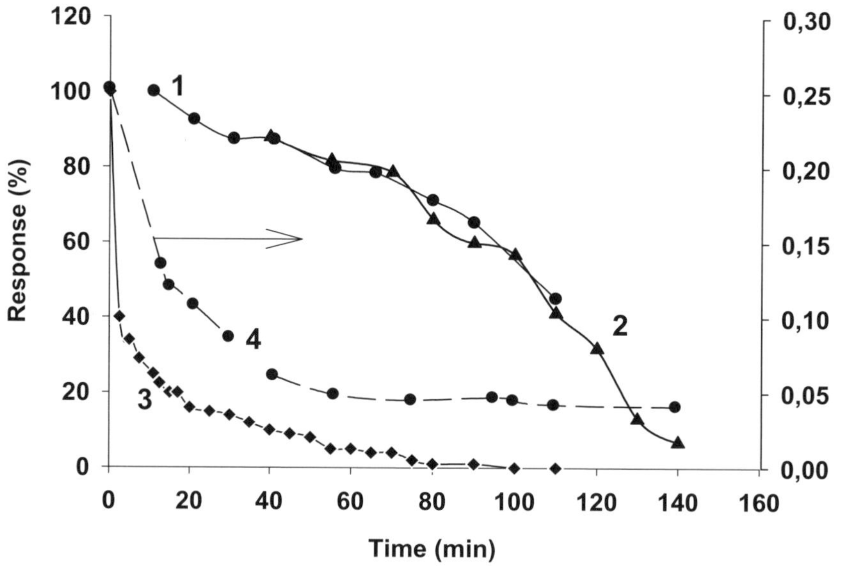

Figure 5. The response of the different glucose biosensors to the consumption of glucose by yeast. Yeast were kept in the 10 ml closed reactor with installed: Clark-type electrochemical oxygen sensor (curve 3), glucose oxidase and hydrogen peroxide electrode-based glucose biosensor (curve 4), and s-GDH and carbon paste–based glucose biosensor (curve 1). Probes of reaction mixture were taken out from the reactor, diluted 50 times with standard buffer and concentration of glucose was measured with glucose analyzer EKSAN-G [19] (curve 2).

As can be seen, during the first two hours, oxygen concentration drops about 5 times (curve 4). The response of the installed glucose biosensor based on glucose oxidase decreases as well, while the consumption of glucose measured by an independent method (curve 2) is not so fast. And only a response of the PQQ-GDH based biosensor correlated with glucose concentration in the reactor. These data illustrate the applicability of these biosensors in microbiological reactors for the process monitoring, as well as in other systems with unstable oxygen concentration, like implantation into blood stream.

A wide substrate specificity of the s-GDH based biosensors allows using them for the determination of some disaccharides in milk and other dairy products [15]. In this case m-GDH based glucose biosensor is used as a reference electrode. Data are presented in Figure 6.

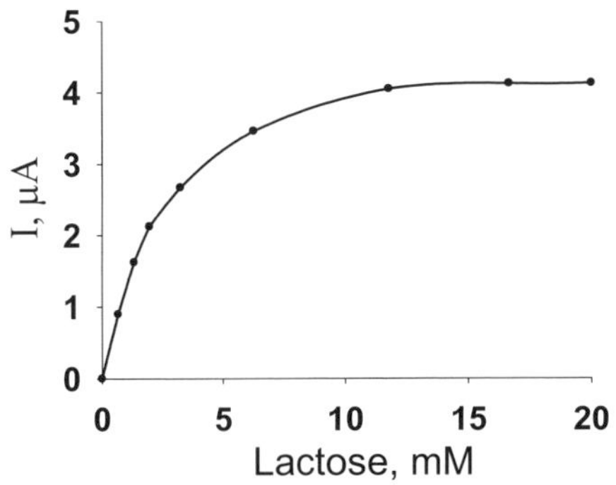

Figure 6. Response of the lactose biosensor to the lactose added to the milk. A working electrode – s-GDH adsorbed on the carbon paste electrode, a reference electrode – m-GDH adsorbed on the carbon paste electrode.

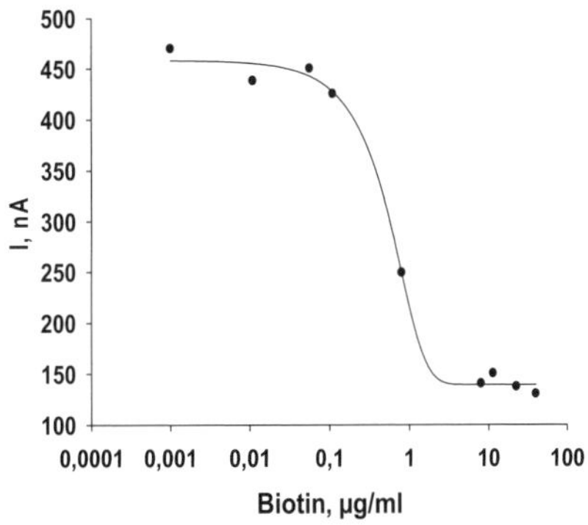

Figure 7. Electrochemical competitive assay for biotin. Neutravidin was adsorbed on the carbon electrode and saturated with the biotinylated s-GDH. Mediator was PMS, potential of 0,4 V was used [20].

High activity of the s-GDH and, thereby, a high sensitivity of the glucose biosensor opens a good possibility to use this enzyme as a marker system in the antibody-antigen analytical systems. Thus, s-GDH was biotinylated with biotinpentafluorophenyl ester. Neutravidine was immobilized on the gold (or carbon paste) electrode. A formed complex between biotinylated enzyme and adsorbed neutravidine was detected electrochemically. Data are shown on Fig. 7 [20].

Alcohol Dehydrogenases in Biosensor Design

m-ADH was purified from *Gluconobacter* sp. 33. The activity of the enzyme was 171 U/ml [21]. 1 mM ethanol was used as a default substrate for all experiments. ADH consists of three sub-units and contains PQQ moiety and four heme-*c* groups. It means, that enzyme can transfer electrons to the acceptors in two ways: (i) directly from PQQ, like PQQ-GDH; (ii) or *via* the hemes chain [5]. A substrate specificity and kinetic parameters of the m-ADH are presented in Table 2. This enzyme prefers three or four carbon atoms containing linear primary alcohols. Methanol is a very bad substrate. ADH can also oxidize longer alcohols such as n-octanol or n-decanol, however, due to the low solubility in water the kinetic parameters were not calculated. Secondary and branched alcohols are unfavorable substrates for this enzyme.

Table 2. Kinetic constants of m-ADH from *Gluconobacter* sp. 33[a]

Substrate	V_{max}[b, c], %	K_M[b], mM	V_{max}/K_M
1-Propanol (0.05-2 mM)	85.5	0.032	2670
Allyl alcohol (0.02-20 mM)	98.6	0.068	1450
1-Butanol (0.02-20 mM)	86.6	0.070	1240
Isoamyl alcohol (0.03-5 mM)	94.2	0.12	772
Ethanol (0.02-20 mM)	100	0.15	667
Isobutyl alcohol	69.9	0.29	243
Ethylene glycol monoethyl ether (0.02-20 mM)	69.3	1.5	46
Propanal (0.005-0.2 mM)	20.9	0.47	44
Propanal (0.2-20 mM)	64.3	2.5	26
2-Propanol (2-20 mM)	86.6	19	4.6
2-Butanol	7.4	73	0.1

[a] Enzyme activity of ADH was measured in 50 mM potassium phosphate buffer (pH 7.3) containing 60 μM PMS, 70 μM DCPIP, and various concentrations of substrate.
[b] Apparent K_M and V_{max} values were determined from double-reciprocal plots.
[c] The enzyme activity with each substrate was expressed relative to a value of 100 for the activity of ADH with ethanol.

An electrochemical regeneration of ADH can be easy performed on gold and platinum electrodes, as well as on the carbon electrodes, directly, or enhanced by soluble or polymeric mediators. Due to this flexibility a number of electrochemical alcohol biosensors can be designed.

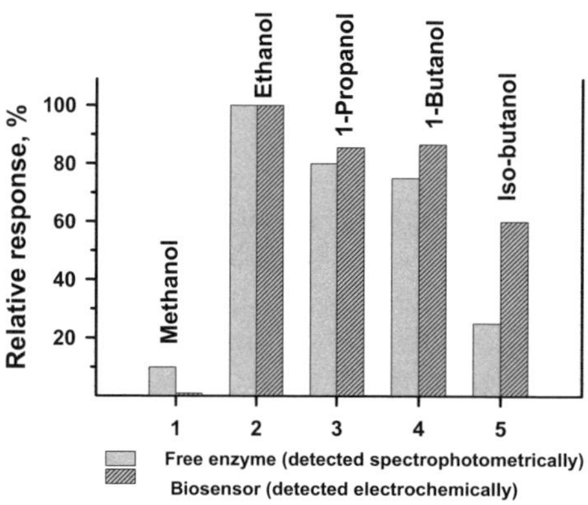

Figure 8. Selectivity of native ADH and biosensor (ADH immobilized in the polypyrrole matrix on Pt electrode [22]).

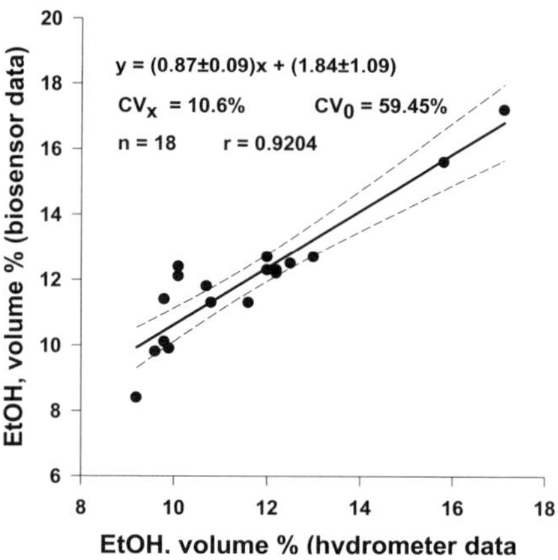

Figure 9. Correlation of ethanol data in red wines obtained with alcohol biosensor (ADH adsorbed on the carbon electrode modified with ferrocenylphenol [23]) and a standard hydrometric method.

Due to the low substrate specificity of the enzyme the alcohol biosensors also possess not very high substrate specificity. On the other hand, a fast electrochemical regeneration of the enzyme makes a substrate binding in the active center as a rate limiting step. Due to heterogeneity of the biosensor and hydrophobic nature of the carbon paste electrode surface the distribution of the substrates near the active center of ADH is influenced. More hydrophobic substrates are concentrated at the electrode surface and change the substrate specificity of the biosensor comparing with the native enzyme in solution (Fig. 8).

Anyway, alcohol biosensor can be successfully applied for the determination of ethanol in wines [23] and other biological materials with predominating ethanol concentration and unstable oxygen concentration.

Glycerol Dehydrogenases in Biosensor Design

Glycerol dehydrogenase (p-Glyc-DH) was purified from the membrane fraction of *Gluconobacter* sp. 33 by the procedures described earlier [24]. The specificity of p-GlycDH was investigated using various mono- and polyalcohols. Results are shown in Table 3. It is worth to note that p-Glyc-DH from *Gluconobacter* sp. 33 prefers the alcohols containing a secondary hydroxyl group and can catalyze oxidation of α-glycerophosphate and alcohols containing a long chain such as n-octanol. The calculated K_m value for glycerol was 0.83 mM.

Table 3. Substrate specificity of p-Glyc-DH from *Gluconobacter* sp. 33

Substrate	Relative activity of GlycDH, %[a]	Substrate	Relative activity of GlycDH, %[a]
Glycerol	100	Isobutanol	4.9
Sorbitol	53.0	Isoamyl alcohol	3.9
Mannitol	35.4	1-Butanol	3.2
2-propanol	29.4	1-Propanol	2.7
1-octanol	26.5[b]	2-Butanol	1.95
Ethylene glycol monoethyl ether	11.8	L-Lactate	1.7
L-α-glycerophosphate	10.3	Ethanol	0.9
Allyl alcohol	8.1	Methanol	0

[a] The p-Glyc-DH activity was measured in 50 mM potassium phosphate buffer, pH 7.3, containing 60 μM PMS, 70 μM DCPIP in the presence of an appropriate substrate (final concentration 20 mM) at 30°C. The activity of p-Glyc-DH in the presence of glycerol was considered as 100
[b] substrate is not miscible in water.

The purified PQQ glycerol dehydrogenase has molecular mass of 72 kDa, and contains one PQQ moiety, which is not tightly bound in the active center. Thereby, the enzyme is rather unstable and lost more than 70% of its activity in 3 days. Immobilization of the enzyme into redox polymers improves the stability of the enzyme, however it is not enough for real application of the biosensors. Often the cell fractions, for example, crude membrane preparations, containing the Glyc-DH activity (m-Glyc-DH) are used for the biosensor design. However, the substrate specificity of such complex is broadened out. Data are presented in Figure 10. Probably, m-GDH also coexists in the same samples.

Glycerol biosensors were designed on the basis of carbon paste modified with 4-ferrocenylphenol and other ferrocene derivatives [25], as well as organic complexes of osmium [26] and p-Glyc-DH or m-Glyc-DH. Biosensors were used for the determination of triglycerides (after hydrolysis with lipases) (Figure 11) and glycerol in wines (Table 4).

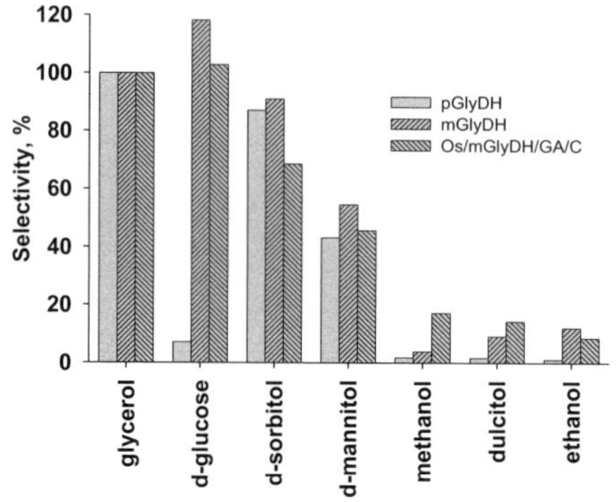

Figure 10. Substrate specificity of p-GlycDH, m-GlycDH and biosensor based on m-GlycDH immobilized on the carbon electrode modified with osmium hydrogel.

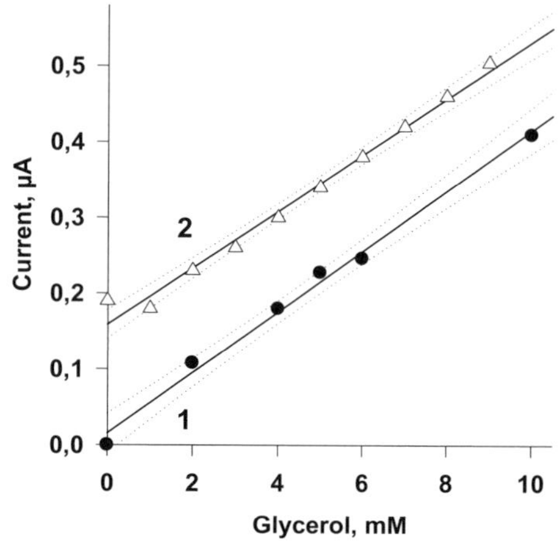

Figure 11. Response of glycerol biosensor (m-Glyc-DH immobilized on carbon electrode, PMS as a mediator) to glycerol (curve 1) and to lipase hydrolyzate of triglycerides (curve 2).

Table 4. Detection of glycerol in wines [25]

	Glycerol, mM	
	Biosensor data*	Reference method data**
Tokaji Aszu (Hyngary)	292±12	240±13
Gluwein (Germany)	177±8	180±6
Kadarka (Germany)	423±16	405±20
Apple wine (Lithuania)	622±3	50±2

* p-Glyc-DH immobilized on the carbon electrode modified with N-(4-hydroxybenzylidene)-4-ferrocenylaniline.

** Spectrophotometric (at 340 nm) method on the basis of commercial test-kit TG GPO-PAP (Roche, France)

References

[1] J.G. Hauge, Glucose dehydrogenase from *Bacterium antiratum*: an enzyme with a novel prosthetic group, *J. Biol. Chem.* **239** (1964), 3603-3639.

[2] J.A. Duine, J. Frank, P.E.J.Verwiel, Structure and activity of the prosthetic group of methanol dehydrogenase, *Eur. J. Biochem.* **108** (1980), 187-192.

[3] K. Matsushita, O. Adachi in: V. Davidson (Ed.). *Principles and Applications of Quinoproteins*, Dekker, New York, (1993), 245-237.

[4] J.A. Duine, The PQQ story, *J.Biosci. Bioeng.***88/3**, (1999), 231-236.

[5] J. Razumiene, R. Meskys, V. Gureviciene, V. Laurinavicius, M. Reshetova, A. Ryabov, 4-Ferrocenylphenol as an electron transfer mediator in PQQ-dependent alcohol and glucose dehydrogenase-catalyzed reactions, *Electrochem. Commun.*, **2**, (2000), 307-311.

[6] R. Le Lagadec, L. Rubio, L. Alexandrova, R.A. Toscano, E. V. Ivanova, R. Meškys, V. Laurinavičius, M. Pfeffer, A. D. Ryabov, Cyclometallated *N,N*-dimethylbenzylamine ruthenium (II) complexes

[Ru($C_6HR^1R^2R^3$-*o*-CH_2NMe_2]PF_6 for bioapplications: synthesis, characterization, crystal structures, redox properties, and reactivity toword PQQ-dependent glucose dehydrogenase. *J. Organometal.Chem.*, **589/25** (2004), 4820-4832.

[7] M. Alkasrawi, I.C. Popescu, V. Laurinavicius, B. Mattiasson, E. Csoregi. A redox hydrogel integrated PQQ-glucose dehydrogenase based glucose electrode. *Anal. Commun.*, **36**, (1999), 395-398.

[8] V. Laurinavicius, B. Kurtinaitienė, V. Liauksminas, A. Ramanavicius, R. Meskys, R. Rudomanskis, T. Skotheim, L. Boguslavsky, Oxygen insensitive glucose biosensor based on PQQ-dependent glucose dehydrogenase, *Anal. Letters,* 32/2, (1999), 299-316.

[9] V. Laurinavičius, B. Kurtinaitienė, V. Liauksminas, B. Puodžiūnaitė, R. Jančienė, L. Kosychova, R. Meškys, A Novel Application of Heterocyclic Compounds for Biosensors Based on NAD, FAD, and PQQ Dependent Oxidoreductases. *Monatshefte fur Chemie,* **130**, (1999), 1269-1281.

[10] V. Laurinavičius, B. Kurtinaitienė, V. Liauksminas, A. Jankauskaitė, R. Šimkus, R. Meškys, L. Boguslavsky, T. Skotheim, S. Tanenbaum. Reagentless biosensor based on PQQ-dependent glucose dehydrogenase and partially hydrolyzed polyarbutin, *Talanta*, **52**, (2000), 485-493.

[11] V. Laurinavičius, J. Razumiene, A. Ramanavicius, A. D. Ryabov, Wiring of PQQ-dehydrogenases. *Biosens. Bioelectronics,* **20,** (2004), 1217-1222.

[12] K. Matsushita, E. Shinagawa, O. Adachi, M. Ameyama, Quinoprotein D-glucose dehydrogenase of the *Acinetobacter calcoaceticus* respiratory chain: membrane-bound and soluble forms are different molecular species, *Biochemistry*, **28**, (1989), 6276-6280.

[13] P. Dokter, J. Frank, J.A. Duine, Purification and characterization of quinoprotein glucose dehydrogenase from *Acinetobacter calcoaceticus L.M.D. 79.41. Biochem. J.,* **239**, (1986), 163-167.

[14] L. Marcinkevičienė, I. Bachmatova, R. Semėnaitė, R. Rudomanskis, G. Bražėnas, R. Meškienė, R., Meškys, Purification and characterization of PQQ-dependent glucose dehydrogenase from *Erwinia* sp. 34-1,. *Biotechnol. Lett.*, **21**, (1999), 187–192.

[15] J. Razumiene, A. Vilkanauskyte, V. Gureviciene, J. Barkauskas, R. Meskys, V. Laurinavičius, Direct electron transfer between PQQ dependent glucose dehydrogenases and carbon electrodes: An approach for electrochemical biosensors, *Electrochimica Acta.* **51**, (2006), 5150-5156.

[16] K. Matsushita, H. Toyama, M. Yamada, O. Adachi, Quinoproteins: structure, function, and biotechnological applications, *Appl. Microbiol. Biotechnol.* **58**, (2002), 13-22.

[17] W. Jin, U. Wolenberger, F.W. Scheller, PQQ as Redox Shuttle for Quinoprotein Glucose Dehydrogenase, *Biol. Chem.* **379**, (1998), 1207-1211.

[18] A.R. Dewanti, J.A. Duine, Ca^{+2}-assisted, direct hydride transfer and rate-determining tautomerization of C5-reduced PQQ to PQQH2 in the oxidation of beta-D-glucose by soluble quinoprotein glucose dehydrogenase, *Biochemistry*, **39**, (2000), 9384-9392.

[19] V. Laurinavicius, J. Kanapienene, A. Dedinaite, B. Kurtinaitiene, Direct Metabolite Determination by Biosensors, *Lecture Notes of the ICB Seminars,* Ed. V. Shumakov and M. Nalecz, Warsaw. (1992), 21-29.

[20] J. Razumiene, A. Vilkanauskyte, V. Gureviciene, I. Bachmatova, L. Marcinkeviciene, V. Laurinavicius, R. Meskys, PQQ-Dependent Glucose Dehydrogenase as Potential Enzymatic Label in Amperometric Avidin-Biotin Systems, *Electroanalysis*, **19/2-3**, (2007), 280-285.

[21] L. Marcinkeviciene, I. Bachmatova, R. Semenaite, R. Rudomanskis, G. Brazenas, R. Meskiene, R. Meskys, Purification and characterization of alcohol dehydrogenase from *Gluconobacter sp 33. Biologija (Vilnius)*, **2**, (1999), 25-30.

[22] A. Ramanavicius, K. Habermuller, J. Razumiene, R. Meskys, L. Marcinkeviciene, I. Bachmatova, E. Csoregi, V. Laurinavicius, W. Schuhmann, An Oxygen-independent ethanol sensor based on quinohemoprotein alcohol dehydrogenase covalently bound to a functionalized polypyrrole film, *Progr. Colloid Polym. Sci.*, **116**, (2000), 143-148.

[23] J. Razumiene, A. Vilkanauskyte, V. Gureviciene, V. Laurinavicius, N.V. Roznyatovskaya, Y.V. Ageeva, M.D. Reshetova, A.D. Ryabov, New bioorganometallic ferrocene derivatives as efficient mediators for glucose and ethanol biosensors based PQQ-dependent dehydrogenases, *J. . Organometal. Chem.*, **668**, (2003), 83-90.

[24] L. Marcinkevičienė, I. Bachmatova, R. Semėnaitė, I. Lapėnaitė, B. Kurtinaitienė, V. Laurinavičius, R. Meškys, Characterization and application of glycerol dehydrogenase from *Gluconobacter* sp. 33, *Biologija (Vilnius)*, 2, (2000), 39-41.

[25] I. Lapenaite, B. Kurtinaitiene, J. Razumiene, V. Laurinavičius, L. Marcinkeviciene, I. Bachmatova, R. Meskys, A. Ramanavicius, Properties and analytical application of PQQ-dependent glycerol dehydrogenase from *Gluconobacter* sp. 33, *Anal. Chim. Acta*, **549**, (2005), 140-150.

[26] M. Niculescu, R. Mieliauskiene, V. Laurinavicius, E. Csoregi, Simultaneous detection of ethanol, glucose and glycerol in wines using pyrroloquinoline quinone-dependent dehydrogenases based biosensors, *Food Chem.,* **82**, (2003), 481-489.

L.M. Lechuga et al. (Eds.)
IOS Press, 2008

Biosensor Detection of Organophosphorous Gases

Jan KREJCI[1], Zuzana GROSMANOVA[1], Dagmar KREJCOVA[1], Petr SKLADAL[2], Bohuslav SAFAR[3]
[1]*BVT Technologies, a. s. Brno, Czech Republic*
[2]*Institute of Biochemistry, Masaryk University, Brno, Czech Republic*
[3]*Military Institute of Protection Brno, Czech Republic*

Abstract. The portable device BioNA for detection of organophosphorous and carbamate substances was developed and tested. The motivation to create such a device was people protection against terrorist attacks by chemical agents (sarin, soman, tabun, VX ...) and improvement of pesticides control in developing countries.
The device consists of analytical block containing the biosensor and diffusion chamber where the toxic chemicals are transferred from sampled air to the circulating solution. The electrical and hydrodynamic circuits are connected together by insertion of the analytical block into the main body of the device.
The main idea behind the device is concept of evaluation of nerve agents presence. It is not measurement of concentration but evaluation of sample toxicity. BioNA device is not still completely finished but it is on such a technical level that it can be reliably decontaminated, it is compact and sufficiently robust to test the biosensor detection possibilities both in laboratory testing chambers and in field trials with real nerve agents spread around the device.

Keywords: nerve agents, biosensor, acetylcholinesterase, portable device, organophosphorous gases

Introduction

Use of chemical weapons by terrorists present high risks against civil people. The preparation of compounds such as sarin require only the basic knowledge of chemistry and with some effort the necessary chemicals can be obtained from common commercial providers. Some types of organophosphorous pesticides exhibit nearly the same toxicity as chemical weapons, nevertheless, they are sold with minimal restrictions. The control of pesticides in developed countries is relatively well regulated – their use is either strictly prohibited or under strict control. However, their use in the developing countries is under completely different conditions, more or less without any control. In many circumstances, they are applied in high amounts, often significantly above the recommended levels [xvii]. This is due to fact that in the case of harvests destroyed by pests the small producer is sometimes in risks of hunger or death.

Such users of organophosphorous toxic compounds have completely different motivation than the people from developed countries. The "safety" has completely different meaning for them compared to the citizens from US and EU, it means to harvest successfully the crop and to survive to next year.

There is no direct economical need for developing a simple, cheap and sensitive detector of organophosphorous compounds from the point of view of the developed countries; the use is strictly regulated and the regulation is commonly maintained. There is however a very strong need to develop such detection means from the following political and human points of view:

- It is necessary to improve the control of pesticides use in the non developed countries (not only pesticides but all toxic compounds present in soil and in drinking water)
- The non developed countries can enable the preparation of toxic compounds for terrorist use due to the insufficient control of the chemical market
- The social differences between developed and no developed regions result in tension which can stimulate the terrorism. Due to the insufficient economical sources, the potential terrorists will focus on chemical and biological weapons as these are cheap.

In this contribution, some initial results from detection of organophosphorous toxic compound by biosensors are presented, focusing mainly on the general technical aspects.

Organophosphorous toxic compounds

The organophosphorous compounds form a subset of nerve agents (NA). The most common toxic warfare agents include **sarin** (**GB**, O-isopropyl methylphosphonofluoridate), **tabun** (**GA**, ethyl N,N-dimethylphosphoramido-cyanidate), **soman**, (**GD**, O-pinacolyl methylphosphonofluoridate), **cyclosarin** (**GF**, cyclohexyl methylphosphonofluoridate), **GV** (P-[2-(dimethylamino)ethyl]-N,N-dimethylphosphonamidic fluoride), **VE** (S-(diethylamino)ethyl O-ethyl ethylphosphonothioate), **VG** (O,O'-Diethyl-S-[2-(diethylamino)ethyl] phosphorothioate, also called **amiton** or **tetram**), **VM** (phosphonothioic acid, methyl-, S-(2-(diethylamino)ethyl) O-ethyl ester) and **VX**, one of the most toxic nerve agents ever synthesized (O-ethyl-S-[2(diisopropylamino)ethyl] methylphosphonothiolate).

These compounds can be considered as first generation nerve agents. The first four were developed by Gerhard Schrader (1903-1990) in Germany between 1936 - 1949. It is ironic that Schrader‘s motivation was humanistic. He tried to develop new insecticides, hoping to make progress in the fight against hunger in the world.

The second generation of chemical weapons is represented by binary chemical weapons. Their principle is based on use of two or more less toxic compounds which are mixed by explosion and the high toxic compound is created by reaction of components. In case of sarin, for example, it is produced by reaction of its precursors methylphosphonyl difluoride and the mixture of isopropyl alcohol and isopropyl amine. The main motivation of the second generation of chemical weapon development laid in the extreme toxicity, production in large amounts and sufficient purity was very difficult. Traces of nerve agents (NA) which can leak have caused numerous fatal accidents among employees. The traces of impurities dramatically increase the autocatalytic decomposition of nerve agents. The production of tabun in Germany during the World War II eventually required sophisticated double wall glass reactors, the space between the walls of the reactor was washed out by air with high pressure, which was collected and filtered.

The third generation of nerve agents was developed in the former Soviet Union considering three strategic aims:

- To circumvent the Chemical Weapons Treaty; the classical chemical weapons were known and their use was prohibited. Thus, the aim was to find new compounds which were not subjected to past treaties.
- To develop new chemical structures which can not be detected by the standard chemical detectors used in NATO troops.
- To defeat NATO chemical personal protection equipment suits for soldiers.

The resulting compounds were known as the **Novichok** agents The existence of these chemical weapons was mentioned in the Moscow News [xvii] in the article by Lev Fedorov and Vil Mirzayanov, the latter author was working on this project and he revealed the government secret. An example of this type of NA is the **A-234** simple unitary agent derived from aconitrile and a common organophosphate pesticide precursor. The compound was prepared as an ultra-fine powder (nanotechnology approach) which provided unique qualities: the nanopowder can bypass most of the chemical weapons detectors commonly used in the modern armies. The nanopowder can, however, firmly attach to skin and the toxic compound directly penetrates into the body.

A number of organophosphorous insecticides such as dichlorvos, malathion and parathion are potent nerve agents and can be used against civilians. Due to different metabolism pathways in insects and mammals, their toxicity is low for humans. However, at high enough doses, their action is completely same as for "real" nerve agents. There is considerable concern about the effects of long-term exposure to such chemicals within farm workers and animals alike. Organophosphate pesticide poisoning is a major cause of disability in many developing countries, and it is often a preferred method of suicide.

Organophosphates (OP) other than the classic nerve agents also exhibit relatively high toxicities, such as amiton (Tetram), Armin, dimefox (Hanane, Terrasytam), paraoxon (E 600), TEPP (Tetron), etc. Consequently, these compounds could be used for terrorist purposes. This class of OP can be described in general as 2-dialkylaminoalkyl-(dialkylamido)-fluorophosphates, structurally similar to the G-compounds (sarin, soman, tabun) and V-compounds (VX and others). Some of them or their analogs were used as pesticides in past namely in the developing countries. Their production for agriculture was ceased due to the high toxicity.

The action of the less toxic organophosphorous compounds can be increased by addition of compounds enhancing skin penetration. O-isopropyl S-2-diisopropylaminoethyl methyl phosphonothiolate has an LD50 in rats of 59.1 μg/kg, but when mixed with dimethylsulphoxide (facilitates skin penetration), the LD50 value is lowered by almost a factor of six (10.1 μg/kg) [xviii].

A similar mechanism as organophosporous compounds is found for carbamates. Their spectrum of toxicity is quite broad from relatively non toxic (carbaryl) to highly toxic compounds comparable with nerve agents (T-1123). The basic mechanism of action is a reversible inhibition of cholinesterases. However, the inhibition is based on carbamoylation of the active centre of cholinesterases. The spontaneous decarbamoylation occurs relatively quickly (approx. 24 h in vivo) and carbamoylated cholinesterases are resistant to effect of reactivators. Therefore the treatment is only symptomatic using preferably atropine. Such difficulties in therapy can potentially be a good reason for military or terrorist use [3]. Furthermore, the plasma cholinesterase activity is also genetically determined and the individuals with decreased cholinesterase activity are more sensitive to myorelaxants, thus more sensitive to nerve agents, too [3].

The existing detection of organophosporous and carbamate compounds is based on the concept of concentration measurement and all decision algorithms are constructed using this concept: it is necessary to determine concentration of the toxic agent and from this information, the conclusion is derived whether the concentration exceeds the limit of either LD50 or LD95 (lethal doses with mortality of either 50% or 95%). The true evaluation of risks using this method needs the knowledge of concentration, chemical composition of the given toxic compound, chemical composition of possibly present toxicity stimulating or suppressing compounds. For all of these components, the toxicological study which relates the concentration to the LD50 or LD95, should be known. In many cases, especially for stimulants, such studies do not exist. Similarly, the integral action of mixed organophosporous and carbamate compounds is usually not considered. It is obvious that neither toxicological data nor chemical structure will be accessible for newly prepared toxic compound. The evaluation of the risk using the concentration-based approach holds a strong historical position which brings some psychological inertia towards alternative more exact approaches. The concept of concentration is very clear from the point of "common sense" but it seems that it is much weaker that the method based on the biosensors technology offering the chance to quantify the real toxicity [xix].

The principle of biosensor nerve agents detection

During the million years of evolution, very effective mechanisms of chemical information evaluation and transmission had been developed. Enzymes, hormones, antibodies, receptors and other biochemical systems may be examples of such mechanisms. These systems are very heterogeneous and very often highly optimized and specialized for certain task by selection during evolution.

The chemical information means not only the knowledge of concentration of the certain substance but also if and how much the sample is toxic. In classical measurement the chemical information is obtained through an elaborate analysis. It mostly requires expensive laboratory equipment, special chemicals and qualified staff. All the individual operational steps require some period of time contributing to the overall costs and time consumption.

The biosensor approach originally offered a completely different motivation. The basic question is: "does the necessary biological system which obtains and processes the chemical information already exist in the nature?" If yes, another question comes: "Could it possibly be isolated and utilized?" The biotechnology is nowadays in such technological level that thousands of biologically active systems can be isolated. The last question appears: "Is it possible to connect the biologically active system to suitable transducer which converts the chemical information acquired by the biological system to electrical signal?" If the answer to the last question is positive we obtain the desired "bioevaluation" of the existing chemical information. If we succeed, then there is high probability to solve the toxicity problem without expensive analytical devices, chemicals, laboratory equipment and qualified staff. The previous statement does not automatically mean that the cheap solution was obtained!

One of the natural systems playing basic role in living bodies is the transmission of neuronal excitement. The basis of this phenomenon is transmission of electrical signal along the neurons. The most important part of this transport of electrical excitation from one neuron to another neuron or to the target cell. If we adopt the biosensor concept, then the information which we are searching for is not the concentration but the "yes / no" type of information - the action potential is transferred to another recipient or not. If transfer becomes blocked, then the toxic compounds are present.

The structure of a biosensor which will be able in the simplest way to evaluate such information is shown in Fig. 1, where its function is compared with the natural synapse function [xx].

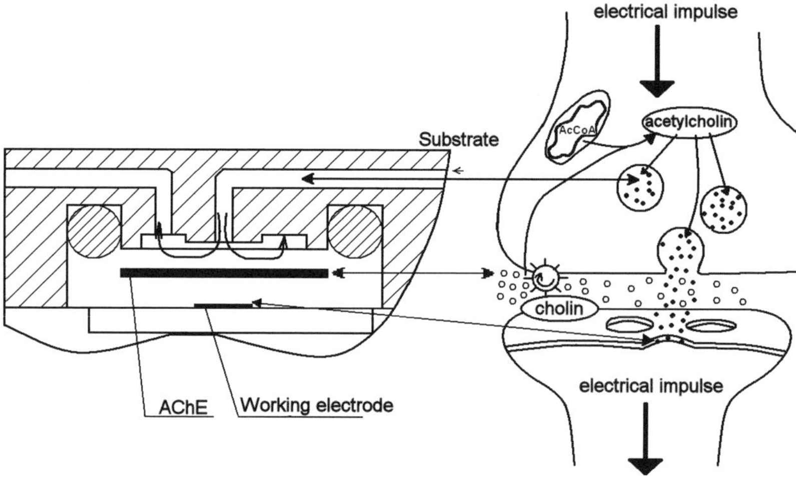

Figure 1. The synapse function (on the right) and its artificial analogy (on the left)

The natural synapses consist of the pre-synaptic part, synaptic gap and postsynaptic part. Acetylcholin (ACh), substrate of enzymatic reaction, is synthesized in the pre-synaptic part and it is concentrated in the vesicules. If the electrical signal reaches the pre-synaptic part, the vesicules release its content to the synaptic gap which is only few nanometers thick. Due to diffusion time of milliseconds ACh reaches the postsynaptic membrane and binds to membrane channel receptors. The ion channels become opened and the membrane is depolarized. The electric action potential propagates to the next cell or neuron. To restore the synapses function it is necessary to close the ion channel and depolarize the post-synaptic membrane. This mechanism is assured by the enzyme acetylcholinesterase (AChE). This enzyme is bound to the postsynaptic membrane and it is also free in the gap. Several oligomeric forms of AChE exist which hydrolyze ACh to acetic acetic acid and choline. Due to changes in concentration of ACh in the synaptic gap, ACh is removed from the receptors and the active ion transport repolarizes the membrane. The choline is resorbed to the presynaptic part, where it is again converted to ACh.

The artificial synapse which is the model of natural synapses is shown in the left part of Fig.1. The transport of ACh from vesicules to synaptic gap is simulated by flow of ACh solution. The reaction of ACh with AChE takes place in the gap where the AChE (from electric eel) is immobilized to the electrochemical sensor.

A question arises how to convert the response of AChE to the electrical signal. If acetylthiocholine is used instead of the natural Ach substrate, the product of enzymatic reaction – thiocholine (TCh) - can be detected by its reaction to disulfide at the electrode at 350 mV (vs. pseudoreference silver electrode):

$$ATCh + H_2O \xrightarrow{AChE} TCh + CH_3COOH$$

$$2TCh \xrightarrow{anode.} disulphide + 2e^-$$

If the non toxic compounds are present then the system provides measurable electric current. It means the biosensor output is related to the activity of AChE. If the signal drops then the AChE must be inhibited. In fact such AChE biosensor was widely described in the literature [xxi, xxii]. Why it is here described as artificial synapsis? The motivation is very simple. The concept of the artificial synapse automatically leads to all possible problems and it motivates further development and improvement of this biosensor concept. The interesting question which immediately follows from the analogy of natural and artificial synapse are listed in short:

- What is the role of different types of AChE in natural synapse?
- Is it possible to simulate the artificial synapse with different types of AChE (at least soluble and immobilized forms)?
- Can some receptors be immobilized into the artificial synapse? If yes then the biosensor will be able to detect additional group of toxic compounds connected with ion channel blocking.

It is also obvious that the artificial synapse will solve some above mentioned problems automatically. If nerve agents will be used together with compounds increasing solubility of AChE inhibitors, then a higher concentration of inhibitor will appear in the fluid coming to the artificial synapse. The inhibition of AChE will be consequently faster and more intensive.

If the immobilized AChE is in contact with inhibitor, then the artificial synapse signal decrease. The relation between signal decrease and concentration of inhibitor was described by the relative inhibition parameter RI [xxiii]:

$$RI = \frac{dI / dt}{I_{SS}}$$

Where dI/dt is the slope of straight line from the point where the inhibition starts and the I_{ss} is the current of the artificial synapse in the moment of inhibition onset. In fact the RI parameter is proportional to the rate constant of binding of inhibitor to the active centre. However, the binding rate of inhibitor to the active centre of AChE represents the chemical information related to the toxic action of NA.. The artificial synapse response is thus proportional to the binding rate of inhibitor to the active centre of the enzyme. The consequences of above mentioned fact are quite important:

- The same signal is obtained for very low concentration of a highly toxic compound and for very high concentration of a compound with low toxic effect.
- The integral toxicity of a mixture of different toxic compounds is "automatically" evaluated. As the real inhibition action is measured the cross-interferences between different toxic compounds are mirrored in the measurement.
- In some cases, the influence of compounds stimulating the toxic effect or decreasing toxic effect are detectable by the artificial synapse, too.

- Very low concentration of inhibitors can be detected.

Late neurotoxic effects

The acute organophosphorous compounds intoxication (complete blocking of synapses excitation) is the principle of NA action as chemical weapons. In case of chronic intoxication the problem is much more complex and many medical phenomena are still not clear. The long term exposure to low concentrations of organophosphorous compounds leads mainly to changes in the immunity system including immuno-supression. This leads to the reduced body resistance to infections and enhanced susceptibility to infections. Sometimes the syndrome of chronic fatigue can be developed. The exposure to low concentration of NA leads also to hormonal disregulation. If young persons are in the contact with low concentrations of NA then the delayed psychomotoric development frequently occurs. The exposure to low concentration of NA leads to decreased immune response to vaccines and to decrease immune response to tumors. The latter leads to higher susceptibility to cancerous diseases. On the other hand, the exposure to low concentration of NA can also stimulate the immune system, resulting in allergies and autoimmune diseases.

Other symptoms are connected with the quality of life. The exposure to low concentration of NA leads to sleeping disorders and nightmares, extreme sensitivity to light, post-traumatic stress disorders, vision problems, both temporary and permanent. The main problem in the above mentioned aspects is represented by the words "chronic" and "long term". Some psychotropic drugs can lead to dependence development even if taken only once. There is no information or clinical studies which determine the time of exposure to low concentrations. A suspicion exists that some symptoms of late neurotoxic effect can develop with a significant delay (weeks or months) after quite short (few hours) exposure to NA without any immediate clinical symptoms.

The problem of late neurotoxic effects was mentioned because the lack of information in this field is caused by the fact that there is still no reliable measurement of very low concentrations of NA. This leads to speculations about alleged intoxications. The Novichok agent, for example, has been suggested as a possible cause of the Gulf War Syndrome. The fact, however, that more common chemical weapons were not identified, suggests that this speculation was more serious than a "conspiracy theory", especially when realizing that the Iraqi chemical weapons stockpiles and capabilities were overestimated or at least remained uncovered.

Generally the biosensor technology can significantly improve the measurement of low concentration of NA and help with the treatment of medical consequences.

Technical solution of biosensor technology for NA measurement

The ideas mentioned in the previous paragraphs were integrated in the portable device BioNA which was developed in the collaboration of Masaryk University, VOP-026 Šternberk, s.p. division VTUO Brno, NBC section and the commercial partner Krejci Engineering. The basic concept of device consists of two functional blocks including the supporting electronics (Fig. 2 - 1) and the analytical block (Fig. 2 - 2). The supporting electronic is a mobile-sized device containing micropumps, microcomputer, accumulator and visible (LED) and audible alarm signalization. The detection part is integrated in the analytical block (2) containing the biosensor, diffusion chamber where the toxic chemicals are transferred from sampled air to the circulating solution. The electrical and hydrodynamic circuits are connected together by insertion of the analytical block into the main body of the device.

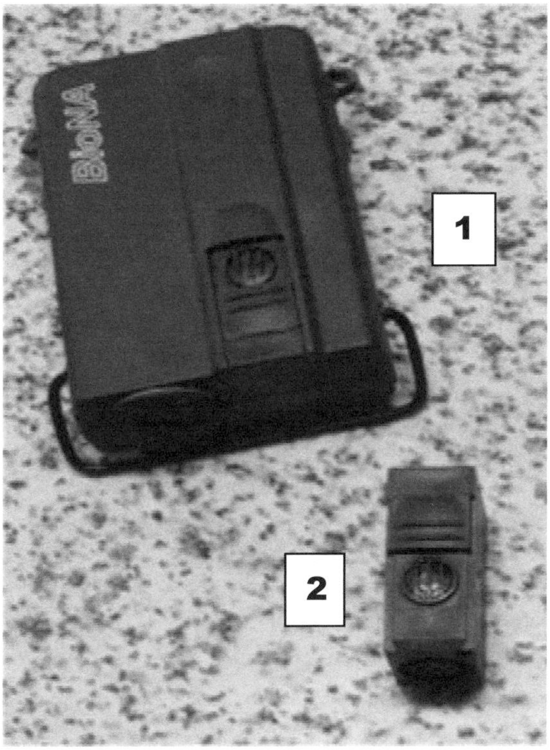

Figure 2. The biosensor-based nerve agent detector BioNA. 1- the main body containing the supporting electronics, 2 – the exchangeable analytical block.

The principle of the analytical block is shown in Fig. 3. It is a disposable unit containing all chemicals needed for the analysis. The main part is the reservoir with substrate solution passing through the diffusion chamber. The volatile agents penetrate into the solution and become transported them to the artificial synapses where their neurotoxicity is tested. The substrate then flows back to the reservoir. The pump which assures the flow is incorporated into the main body and it is connected via a special connector assurring reliable electrical and hydrodynamic connections. The connector also contains the mechanism which opens the reservoir; when the analytical block is inserted to the device, all hydrodynamic paths are connected and the analytical block becomes activated. All hydrodynamic paths are kept dry during storage and the analytical block can be stored 3 months.

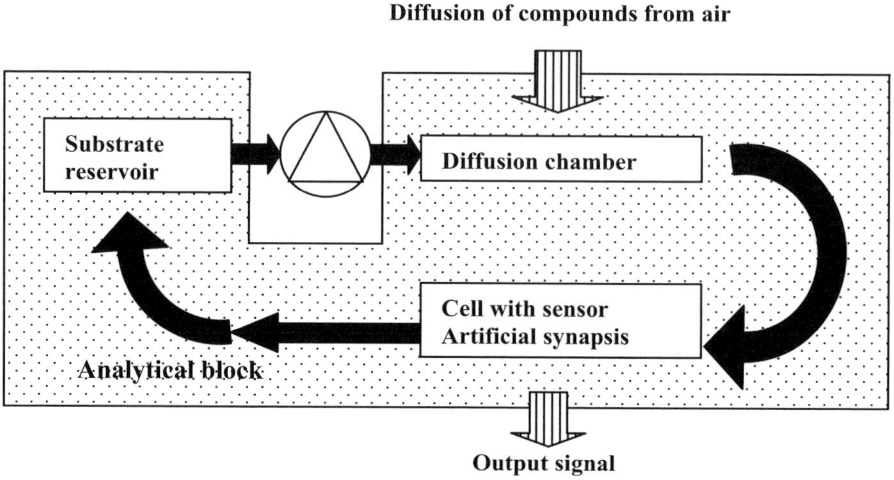

Figure 3. The principle of function of the analytical block.

Figure 4. The combined connector connecting the electric and hydrodynamic circuits of analytical block. The connector embedded in the main part of the device is shown at left, the corresponding counter-connector in the analytical block is shown at right.

Results and discussion

The concept of analysis of information contained in the measurement is the most important idea behind the biosensor-based assay. To emphasize this idea, one should consider three independent measurements of a sample containing tabun and some unknown agent of V type. The analysis is carried out using gas chromatography (GC), ion mobility spectrometry and the biosensor. In case of GC it is necessary to collect the sample and analyze it. The resulting information says that the sample contained tabun and an unknown compound probably belonging to organophosphates. As the sample contains the unknown chemical it is impossible to predict anything about toxicity of the sample. It is not possible to decide if the toxicity is caused by tabun or by the unknown agent or if the toxic action is enhanced by the common action of both agents. The obtained information is that the sample contains the concentration of tabun at some mg/l and the concentration of unknown chemical at some mg/l.

If the ion mobility spectrometry (IMS) is used instead, the resulting signal will be indicate that the sample contains some organic chemicals which have the same mobility as toxic compounds for which the detector was pre-programmed. Such information is very important in case of military use in field where only possible organic chemical can be the chemical weapons of enemy. However, in protection of building against terrorists attack, for example, there are many chemicals which can trigger the false positive signal. An example might be e.g. compounds applied for preservation of flowers. In fact, the obtained information is that the sample contains the chemicals which have the same mobility as group of mobility spectra stored in the device memory.

In case of biosensor measurement, the sample is in contact with the opened window of the diffusion chamber where the volatile agents penetrates to the liquid. The liquid carries the dissolved toxic compounds to the artificial synapses where the signal is obtained which is proportional to the range of AChE inhibition. The diffusion chamber can simulate the penetration of sample through skin. If the biosensor system is calibrated with sarin then its output signal is proportional to the concentration of sarin of such magnitude that will exhibit the same toxic action as the measured sample. The question is what information is the most valuable, what information is the most precise, what information is most costly and what information is obtained faster. Our estimation of answer to such questions is summarized in Table 1.

The next aspect of the biosensor measurement is its high sensitivity. This fact is particularly important in protection against terrorist attacks. It can be supposed that in such attacks against civilians they will not be prepared for such action. The time in minutes may be decisive to initiate rescue actions efficiently. It is necessary to decide two situations. The concentration rises immediately. In this case the faster is the detector reaction the higher probability of saving persons. A fast detector is needed. In the second case the toxic agent diffuses inside the building. In this case the detection of lower concentrations prolongs the time in which the rescue action can be done. The sensitive detector is needed. The overall situation is summarized in Fig. 5 (adapted from Safar [xxiv]). The time on the y axis means the time when 50% of persons with contact with NA will demonstrate the first symptoms of NA poisoning. It is seen that the limit of detection of biological methods based on inhibition of AChE (CHP-71, GSP-11, BioNA) has such low limit detection that the people can survive hours in the contaminated area. The fast detectors (RAID, M-90, CAM) mostly based on ion mobility spectrometry have lower detection limit which enables to survive only a few minutes. The advantage of the BioNA concept lies in the fact that the device is fully autonomous and it does not need any attention during its function. CHP-71 is a manually operated device and GSP-11 is automated but rather big device intended for mobile laboratories.

Table 1. Comparison of different methods for nerve agents analysis

Method	Information			
	value with respect of toxicity	precision	cost	response time
GC	medium	high	high	low
IMS	low	medium	high	high
Biosensor	high	low	low	medium

Comments:

- The biosensor measurement precision is low due to fact that the measurement is based on a biological object which is variable and due to fact that many technological problems of biosensor technology are still not appropriately solved (e.g. immobilization of AChE with sufficient long term stability).
- Time of response GC is low – the sample must be prepared and a complicated laboratory device is used.

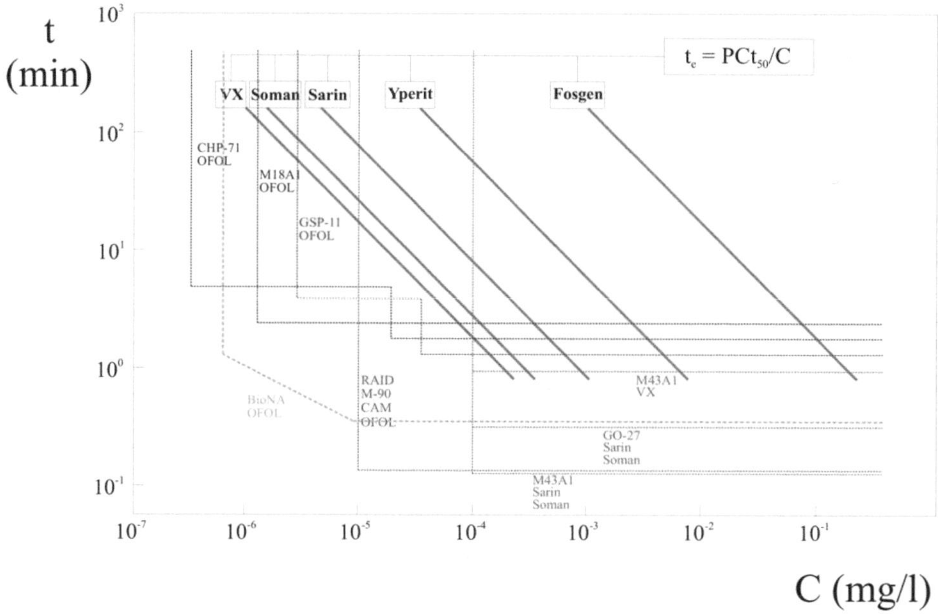

Figure 5. The dependence of time when the first symptoms are present at 50% of persons exposed to different toxic agents in dependence on the agent concentration.

The calibration curve of BioNa for sarin is presented in Fig. 6. The measurement was done with the whole BioNA analyser being inserted to the chamber containing the given concentration of sarin.

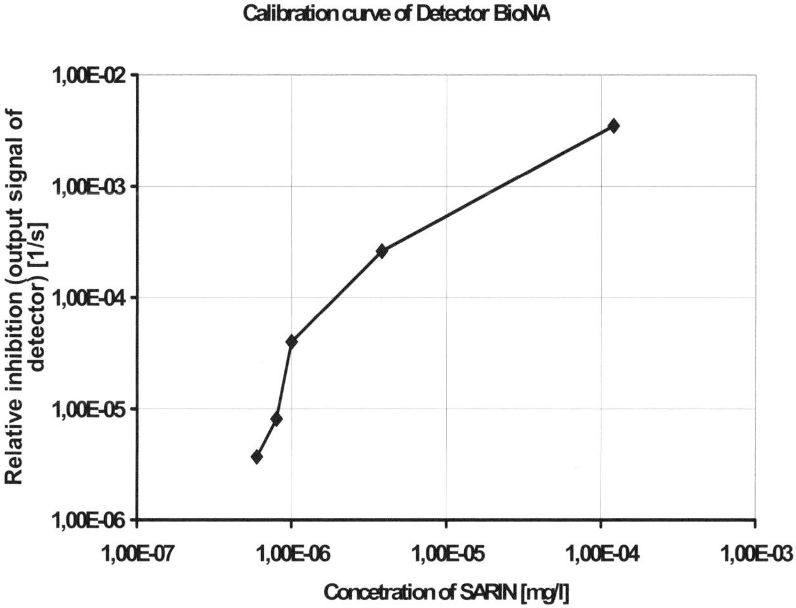

Figure 6. Calibration curve of BioNA for sarin.

If the person is subjected to concentration of 10^{-6} mg/l, then the time till the first symptoms of intoxication will be from 30 to 150 minutes. If the concentration reaches 10^{-5} mg/l, then the person survives only few minutes. The comparison of Figs. 5 and 6 proves that the biosensor enables to detect the concentration significantly lower than other detecting devices. This means that the affected person can carry out some protective actions in due time.

The long term stability of BioNA measurement is shown in Fig. 7. The results of a 24 hours test are shown. It is obvious that the signal of BioNA slightly raises with time for two reasons. The first one is the increase of concentration of the product thiocholine which becomes gradually accumulated in the substrate reservoir, the second one is the slow swelling of the bioactive membrane. After 24 hours the device was brought into contact with toxic compound (1 µmol of neostigmine) which caused the decrease of signal. The response time was 20 s to obtain 50% inhibition. This example demonstrates the possibility of a 24 hours continuous monitoring period using the biosensor measurement. It is worth to mention that the toxic compounds are accumulated in the internal reservoir of the analytical block. If a very long measurement is employed, very low concentrations can be detected. This measurement was still not finished. The accumulation of the toxic compounds enables the evaluation of the resulting integral action. It is equivalent to a toxicity dose.

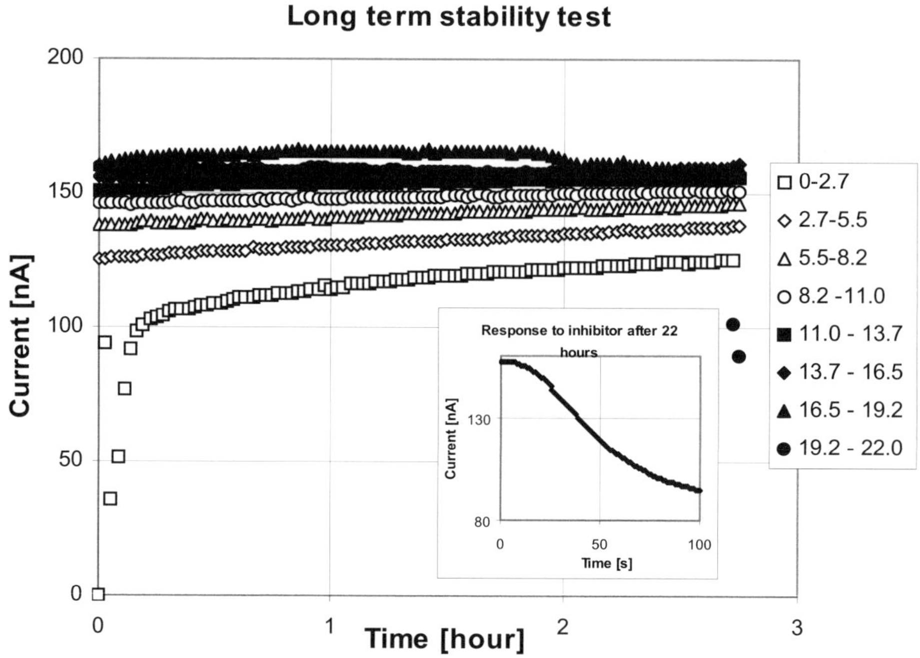

Figure 7. Long term stability of BioNA measurement. 22 hour profile was finished by contact with the AChE inhibitor neostigmin.

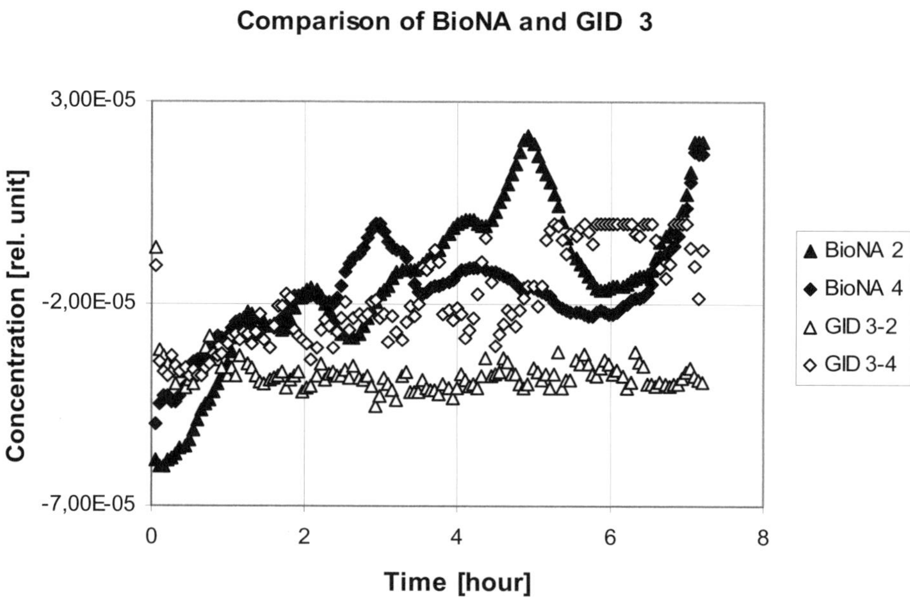

Figure 8. The comparison of BioNA measurement and GID 3 measurement.

Fig. 8 shows the comparison of BioNA measurement with the standard device GID 3. The measurement was carried out in field in the testing corridor of VOP – 026 Štenberk, s.p. division VTUO Brno, NBC section. The devices GID 3 and BioNA were in different positions above the concrete testing area where sarin was applied. Due to diffusion and evaporation sarin penetrates to higher levels and also the vapours become diluted by air movement. The GID-3 and BioNA devices were in different positions above the concrete testing area where sarin was applied. Due to diffusion and evaporation sarin penetrates to higher levels and also the vapors become diluted by air movement. The GID-3 device analyzed the sample which was transferred to the device by a forced air flow. Active surface of the BioNA device was only in a passive contact with the surrounding air - no forced flow around its active surface was assured. The GID-3 in position 2 and BioNA in position 2 units were close to the contaminated surface. The GID-3 in position 4 and BioNA in position 4 units were in a height about 1 m above the contaminated surface. The signals of both devices were normalized with respect of the maximum measured concentration. It is seen that in the case of the BioNA device in position 2 which was measuring very high concentration, the enzymatic activity decreased very fast.. After approximately 3 hours the enzymatic activity was exhausted and the signal dropped to zero. The fluctuations of signal were caused by the ambient air movement which changed the mass transfer between air and the diffusion chamber of BioNA.

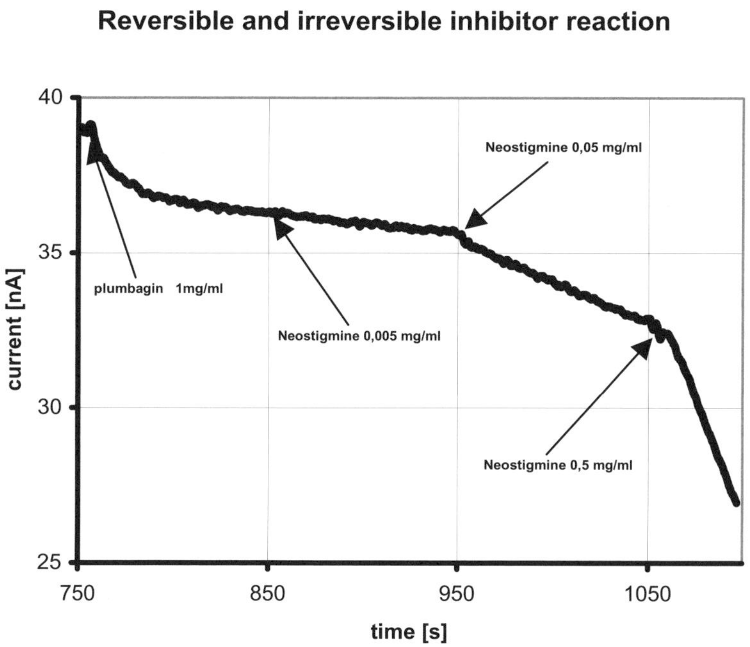

Figure 9. The difference between response of reversible (the first decrease) and irreversible inhibitor (neostigmine)

In case of BioNA in position 4 which measured lower concentrations the agreement between both devices is relatively very good in time form 0 to 5000 s (1.5 hour). Then the results of both devices start to differ. Both devices have higher fluctuations of the measured signals. It is supposed that the reason for this lies in fact that the measurement is realized 1 m above the contaminated area and the influence of air movement is significantly higher than in case of measurement closer to the contaminated area.

Fig. 9 demonstrates the detailed analysis of the BioNA signal. The first decrease on the signal trace (near 750 s) corresponds to the response of AChE after contact with a reversible inhibitor (extract from a plant). Another explanation is that the plant extract might be quickly hydrolyzed in a measured solution. The structure or any other information about its chemical composition is not known. The inhibition capability was proven independently using the Elman method. The second decrease (around 950 s) is caused by an irreversible inhibitor (neostigmine). The example of this experiment shows that not only the toxicity can be detected by biosensor based device but also some information about the mechanism of toxic action can be obtained.

Conclusions

We consider as very important to discuss the relation of above mentioned results with respect to biosensor technology and application of the BioNA type device on market. A significant difference exists between prove of the principle and technical solution suitable for real use. The prove of principle of detection of NA by inhibition of AChE is more than 60 years old. In our case, it was successfully demonstrated that this principle can be integrated in the mobile-size device. To be critical, some experiments were carried out only in few repetitions and some details of the technical solution were not completely optimized.

On the other hand the technical solution should solve such questions as device reliability, robustness of solution, precision, and exactness. All presented results need to be expressed with errors and sufficient statistical evaluation. This task is extremely time consuming and costs significant resources; it was not completely finished in our case. This task is also very laborious and only few scientists admit that the technical solution could need wide intellectual capabilities and knowledge. The technical results are hard to get published so the scientific interest usually turns out to other “more interesting” problems.

Furthermore, we consider as quite important two results. The first one is the change of the concept of evaluation of presence of NA from measurement of concentration to evaluation of information which is obtained from the “biodevice”. The second concept is the key idea of BioNA as a technical device. The core of problems lies in fact that the BioNA device is not still completely finished but it is on such a technical level that it can be reliably decontaminated, it is compact and sufficiently robust to test the biosensor detection possibilities both in laboratory testing chambers and in field trials with real nerve agents spread around the device. This fact creates the real base for the next development of this technology towards to very effective mobile sized device which can serve for personal protection. Another very important fact is that the device is fully autonomous and automated. The user has free hands and he is for 24 hour protected. In case of the common detection tubes, the user must operate each hour or other time period the device and he must focus only on the analysis.

We believe that the presented concept has great potential in the future namely with the fact that many technical solutions used in BioNA can be also applied in the detection of other potential threads such as biological agents.

Acknowledgements

We express the acknowledgement to Czech Ministry of Defence, which support the development of BioNA in Past. To the VOP –026 Šternberk, s.p. division VTUO Brno, section NBC which coordinated the development and where all test with NA was carried out. We are also graetfull to Oldriska Mojzisova, who finalized the manuscript.

Reference

[xvii] L. Fedorov, V. Mirzayanov, A Poisoned Policy, *Moscow News* weekly No. 39, 1992.

[xviii] J. Bajgar, Some Toxic Chemicals as Potential Chemical Warfare Agents - The Threat for the Future?, *ASA Newsletter* **98-6** (1998)

[xix] P. Skladal, M. Fiala, J. Krejci, Detection of pesticides in the environment using biosensors based on cholinesterases. *Int. J. Environ. Anal. Chem.* **65** (1996), 139-148.

[xx] Z. Grosmanova, J. Krejci, J. Tynek, P. Cuhra, S. Barsova, Comparison of biosensoric and chromatographic methods for the detection of pesticides, Int. J. Environ. Anal. Chem. 85 (2005), 885-893.

[xxi] P. Skladal, Cholinesterase-based biosensors for detection of pesticides, *Food Technol. Biotechnol.* **34** (1996), 43-49.

[xxii] S. Andreescu, J.L. Marty, Twenty years research in cholinesterase biosensors: From basic research to practical applications, *Biomol. Eng.* **23** (2006), 1-15.

[xxiii] P. Skladal, Determination of organophosphate and carbamate pesticides using a cobalt phthalocyanine-modified carbon paste electrode and a cholinesterase enzyme membrane, *Anal. Chim. Acta* **252** (1991) 11-15.

[xxiv] B. Safar, Poison substances analytical determination by laser optoacoustic spectrometer, *doctoral thesis*, Military Academy of Antonin Zapotocky (1987)

L.M. Lechuga et al. (Eds.)
IOS Press, 2008

Author Index